Bearings and Lubrication

A Mechanical Designers' Workbook

Other Books in The McGraw-Hill Mechanical Designers' Workbook Series

MACHINE DESIGN FUNDAMENTALS: A Mechanical Designers' Workbook

DISTORTION AND STRESS: A Mechanical Designers' Workbook

CORROSION AND WEAR: A Mechanical Designers' Workbook

FASTENING, JOINING, AND CONNECTING: A Mechanical Designers' Workbook

GEARING: A Mechanical Designers' Workbook

MECHANISMS: A Mechanical Designers' Workbook

POWER TRANSMISSION ELEMENTS: A Mechanical Designers' Workbook

Bearings and Lubrication

A Mechanical Designers' Workbook

Editors in Chief

Joseph E. Shigley

Professor Emeritus
The University of Michigan
Ann Arbor, Michigan

Charles R. Mischke

Professor of Mechanical Engineering
Iowa State University
Ames, Iowa

McGraw-Hill Publishing Company

New York St. Louis San Francisco Auckland Bogotá
Caracas Hamburg Lisbon London Madrid Mexico
Milan Montreal New Delhi Oklahoma City
Paris San Juan São Paulo Singapore
Sydney Tokyo Toronto

Library of Congress Cataloging-in-Publication Data

Main entry under title:

Standard handbook of machine design.

Includes index.
1. Machinery—Design—Handbooks, manuals, etc.
I. Shigley, Joseph Edward. II. Mischke, Charles R.
TJ230.S8235 1986 621.8′15 85-17079
ISBN 0-07-056892-8
ISBN 0-07-056928-2 (workbook)

1234567890 KGP/KGP 8965432109

ISBN 0-07-056928-2

The editors for this book were Robert Hauserman and Scott Amerman, and the production supervisor was Dianne Walber. It was set in Times Roman by Techna Type.

Printed and bound by The Kingsport Press.

In Loving Memory of
Opal Shigley

CONTENTS

CONTRIBUTORS

Daniel E. Czernik, *Director of Product Engineering,* Fel-Pro Inc., Skokie, Illinois

R. Bruce Hopkins, The Hopkins Engineering Co., Cedar Falls, Iowa

Theo G. Keith, Jr., *Professor and Chairman of Mechanical Engineering,* University of Toledo, Ohio

A. R. Lansdown, *Director,* Swansea Tribology Centre, University College of Swansea, United Kingdom

Charles R. Mischke, *Professor of Mechanical Engineering,* Iowa State University, Ames, Iowa

ABOUT THE EDITORS

Joseph E. Shigley has had a long and distinguished career both as an educator and as a consultant in machine design and mechanical engineering. He was a professor of mechanical engineering at the University of Michigan for 21 years. He is a Fellow of the American Society of Mechanical Engineers. He received the ASME Mechanisms Committee Award in 1974, the Worchester Reed Warner Medal in 1977, and the Machine Design Award in 1985. He is the author or co-author of eight McGraw-Hill books.

Charles R. Mischke is Professor of Mechanical Engineering at Iowa State University, a consultant to industry, contributor of many technical papers, author or co-author of five books, and co-editor of the *Standard Handbook of Machine Design.* He received the Ralph Teeter Award of the Society of Automotive Engineers in 1977, and the University's Outstanding Teacher Award in 1980. A Fellow of the American Society of Mechanical Engineers (ASME), he also serves on its Reliability, Stress Analysis, and Failure Prevention Executive Committee.

PREFACE

There is no shortage of good textbooks treating the subject of machine design and related topics of study. But the beginning designer quickly learns that there is a great deal more to successful design than is presented in textbooks or taught in technical schools or colleges. A handbook connects formal education and the practice of design engineering by including the general knowledge required by every machine designer.

Much of the practicing designer's daily informational needs are satisfied in various pamphlets or brochures such as are published by the various standards organizations as well as manufacturers of various components used in design. Other sources include research papers, design magazines, and corporate publications concerned with specific products. More often than not, however, a visit to a design library or to a file cabinet will reveal that a specific publication is on loan, lost, or out of date. A handbook is intended to serve such needs quickly and immediately by giving the designer authoritative, up-to-date, understandable, and informative answers to the hundreds of such questions that arise every day in the work of a designer.

The *Standard Handbook of Machine Design** was written for working designers, and its place is on their desks, not on their bookshelves, for it contains a great many formulas, tables, charts, and graphs, many in condensed form. These are intended to give quick answers to the many questions that seem to arise constantly.

The *Mechanical Designers' Workbook* series consists of eight volumes, each containing a group of related topics selected from the *Standard Handbook of Machine Design*. Limiting each workbook to a single subject area of machine design makes it possible to create a thin, convenient volume bound in such a manner as to open flat and provide an opportunity to enter notes, references, graphs, equations, standard corporate practices, and other useful data. In fact, each chapter in every workbook contains gridded pages located in critical sections for this specific purpose. This flat-opening workbook is easier to use on the designer's work space and will save much wear and tear on the source handbook.

Chapter 1 of *Bearings and Lubrication* deals with the lubricant itself and with the systems for applying it. The characteristics, limitations, and advantages of liquid, solid, and gas lubricants and the greases are discussed in detail and important characteristics displayed and compared in graphic form. Feeding, storage, and lubricant deterioration problems are all probed.

When machine parts are assembled it is often necessary to employ gaskets and seals to isolate various chambers and spaces to prevent leakage and maintain pressure and lubricant supply. Chapter 2 deals with this subject and shows the designer how to isolate such volumes. Types, properties, limitations, and advantages of these devices are brought out, together with installation and maintenance procedures.

So-called antifriction bearings consisting of rolling balls and rollers are discussed in Chapter 3. Though most bearing manufacturers provide excellent manuals to aid the designer,

*By Joseph E. Shigley and Charles R. Mischke, Coeditors-in-Chief, McGraw-Hill Publishing Company, New York, 1986.

they don't often tell the whole story. This chapter includes the basics of rolling-bearing selection with a detailed treatment of the interrelationships between load, life, and reliability, all independent of the various manufacturers' particular procedures.

This volume concludes with the treatment of the selection, lubrication, and performance of journal bearings in Chapter 4. Charts, graphs, and worked-out examples are used generously to show how the performance of such bearings can be predicted. Hydrostatic and gas-lubricated bearings are also discussed.

Most of the artwork has been competently prepared and supervised by Mr. Gary Roys of Madrid, Iowa, to whom the editors are indebted.

Care has been exercised to avoid error. The editors will appreciate being informed of errors discovered so that they may be eliminated in future printings.

JOSEPH E. SHIGLEY
CHARLES R. MISCHKE

chapter 1
LUBRICATION

A. R. LANSDOWN, M. Sc., Ph. D.
Director, Swansea Tribology Centre
University College of Swansea
Swansea, United Kingdom

1-1 FUNCTIONS AND TYPES OF LUBRICANT

Whenever relative movement takes place between two surfaces in contact, there will be resistance to movement. This resistance is called the *frictional force,* or simply *friction.* Where this situation exists, it is often desirable to reduce, control, or modify the friction.

Broadly speaking, any process by which the friction in a moving contact is reduced may be described as *lubrication.* Traditionally this description has presented no problems. Friction reduction was obtained by introducing a solid or liquid material, called a *lubricant,* into the contact, so that the surfaces in relative motion were separated by a film of the lubricant. Lubricants consisted of a relatively few types of material, such as natural or mineral oils, graphite, molybdenum disulfide, and talc; and the relationship between lubricants and the process of lubrication was clear and unambiguous.

Recent technological developments have confused this previously clear picture. Friction reduction may now be provided by liquids, solids, or gases or by physical or chemical modification of the surfaces themselves. Alternatively, the sliding components may be manufactured from a material which is itself designed to reduce friction or within which a lubricant has been uniformly or nonuniformly dispersed. Such systems are sometimes described as "unlubricated," but this is clearly a matter of terminology. The system may be unconventionally lubricated, but it is certainly not unlubricated.

On the other hand, lubrication may be used to modify friction but not specifically to reduce it. Certain composite brake materials may incorporate graphite or molybdenum disulfide, whose presence is designed to ensure steady or consistent levels of friction. The additives are clearly lubricants, and it would be pedantic to assert that their use in brake materials is not lubrication.

This introduction is intended only to generate an open-minded approach to the processes of lubrication and to the selection of lubricants. In practice, the vast majority of systems are still lubricated by conventional oils or greases or by equally ancient but less conventional solid lubricants. It is when some aspect of the system makes the use of these simple lubricants difficult or unsatisfactory that the wider interpretation of lubrication may offer solutions. In addition to their primary function of reducing or controlling friction, lubricants are usually expected to reduce wear and perhaps also to reduce heat or corrosion.

In terms of volume, the most important types of lubricant are still the liquids (oils) and semiliquids (greases). Solid lubricants have been rapidly increasing in importance since about 1950, especially for environmental conditions which are too severe for oils and greases. Gases can be used as lubricants in much the same way as liquids, but as is explained later, the low viscosities of gases increase the difficulties of bearing design and construction.

1-2 SELECTION OF LUBRICANT TYPE

A useful first principle in selecting a type of lubrication is to choose the simplest technique which will work satisfactorily. In very many cases this will consist of a small quantity of oil or grease which is inserted in the component on initial assembly and which is almost never replaced or refilled. Typical examples are door locks, hinges, car-window winders, switches, clocks, and watches.

This simple system is likely to be unsatisfactory if the loads or speeds are high or if the service life is long and continuous. Then it becomes necessary to choose the lubricant with care and often to use a replenishment system.

The two main factors in selecting the type of lubricant are the speed and load. If the speed is high, then the amount of frictional heating tends to be high, and low-viscosity lubricants will give lower viscous friction and better heat transfer. If the loads are high, then low-viscosity lubricants will tend to be expelled from the contact. This situation is summarized in Fig. 1-1.

It is difficult to give precise guidance about the load and speed limits for the various lubricant types, because of the effects of geometry, environment, and variations within each type; but Fig. 1-2 gives some approximate limits.

Some other property of the system will sometimes restrict the choice of lubricant type. For example, in watches or instrument mechanisms, any lubricant type could meet the load and speed requirements, but because of the need for low friction it is normal to use a very low-viscosity oil. However, for open gears, wire ropes, or chains the major problem is to prevent the lubricant from being thrown off the moving parts, and it is necessary to use a "tacky" bituminous oil or grease having special adhesive properties.

In an existing system the geometry may restrict the choice of lubricant type. Thus, an unsealed rolling bearing may have to be lubricated with grease because oil would not be retained in the bearing. But where the lubrication requirements are difficult or particularly important, it will usually be essential to first choose the lubricant type and then design a suitable system for that lubricant. Some very expensive mistakes have been made, even in high technology such as aerospace engineering, where systems have been designed and built but could not be lubricated.

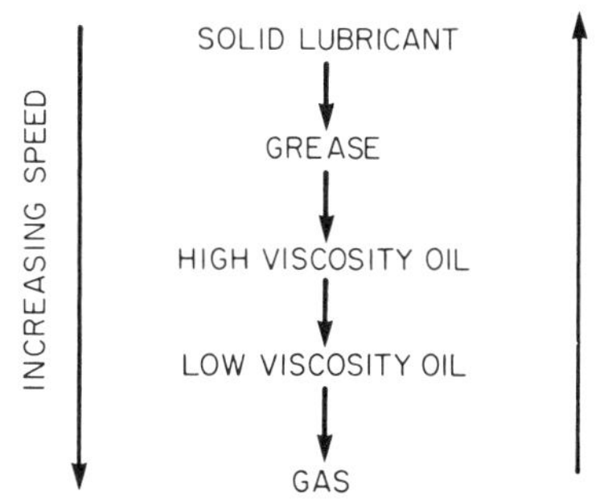

FIG. 1-1 Effect of speed and load on choice of lubricant type. *(From Ref. [1-1].)*

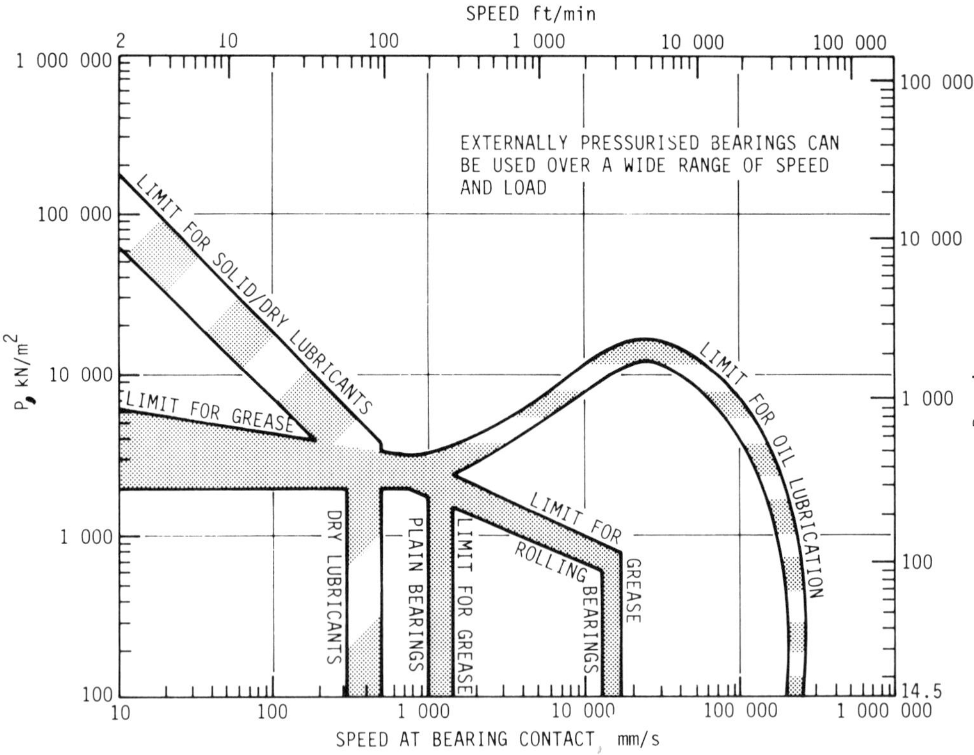

FIG. 1-2 Speed and load limitations for different types of lubricant. *(From Ref.* [*1-2*].*)*

1-3 LIQUID LUBRICANTS: PRINCIPLES AND REQUIREMENTS

The most important single property of a liquid lubricant is its viscosity. Figure 1-3 shows how the viscosity of the lubricant affects the nature and quality of the lubrication. This figure is often called a *Stribeck curve,* although there seems to be some doubt as to whether Stribeck used the diagram in the form shown.

The expression $\eta N/P$ is known as the *Sommerfeld number,* in which η is the lubricant viscosity, N represents the relative speed of movement between the counterfaces of the bearing, and P is the mean pressure or specific load supported by the bearing. Of these three factors, only the viscosity is a property of the lubricant. And if N and P are held constant, the figure shows directly the relationship between the coefficient of friction μ and the lubricant viscosity η.

The graph can be conveniently divided into three zones. In zone 3 the bearing surfaces are fully separated by a thick film of the liquid lubricant. This is, therefore, the zone of *thick-film* or *hydrodynamic lubrication,* and the friction is entirely viscous friction caused by mechanical shearing of the liquid film. There is no contact between the interacting surfaces and therefore virtually no wear.

As the viscosity decreases in zone 3, the thickness of the liquid film also decreases until at the point C it is only just sufficient to ensure complete separation of the surfaces. Further reduction in viscosity, and therefore in film thickness, results in occasional contact between asperities on the surfaces. The relatively high friction in asperity contacts offsets the continuing reduction in viscous friction, so that at point B the friction is roughly equal to that at C.

Point C is the ideal point to provide zero wear with almost minimum friction,

Notes ▪ Drawings ▪ Ideas

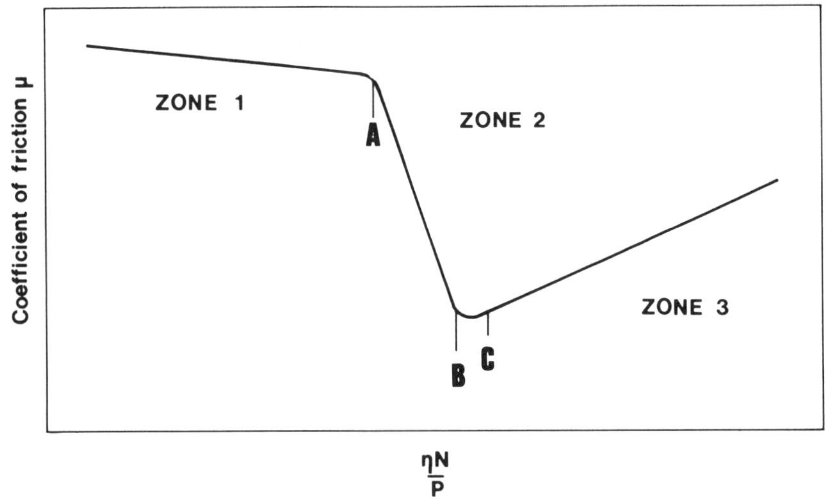

FIG. 1-3 Effect of viscosity on lubrication.

but in practice the design target will be slightly to the right of C, to provide a safety margin.

With further reduction in viscosity from point B, an increasing proportion of the load is carried by asperity contact, and the friction increases rapidly to point A. At this point the whole of the bearing load is being carried by asperity contact, and further viscosity reduction has only a very slight effect on friction.

Zone 1, to the left of point A, is the zone of *boundry lubrication.* In this zone, chemical and physical properties of the lubricant other than its bulk viscosity control the quality of the lubrication, and these properties are described in Sec. 1-5.

Zone 2, between points A and B, is the zone of mixed lubrication in which the load is carried partly by the film of liquid lubricant and partly by asperity interaction. The proportion carried by asperity interaction decreases from 100 percent at A to 0 percent at C.

Strictly speaking, Fig. 1-3 relates to a plain journal bearing and then N usually refers to the rotational speed. Similar patterns arise with other bearing geometries in which some form of hydrodynamic oil film can occur.

The relationship between viscosity and oil-film thickness is given by the Reynolds equation, which can be written as follows:

$$\frac{\delta}{\delta x}\left(h^3 \frac{\delta P}{\delta x}\right) + \frac{\delta}{\delta z}\left(h^3 \frac{\delta P}{\delta z}\right) = \eta\left(6U\frac{\delta h}{\delta x} + 6h\frac{\delta U}{\delta x} + 12V\right)$$

where h = lubricant-film thickness
P = pressure
x, z = coordinates
U, V = speeds in directions x and z

Fuller details of the influence of lubricant viscosity on plain journal bearings are given in Chap. 4.

In nonconformal lubricated systems such as rolling bearings and gears, the relationship between lubricant viscosity and film thickness is complicated by two additional effects—the elastic deformation of the interacting surfaces and the increase in lubricant viscosity as a result of high pressure. The lubrication regime is then known as *elastohydrodynamic* and is described mathematically by various equations.

For roller bearings a typical equation is the Dowson-Higginson equation

$$h_{\min} = \frac{2.65(\eta_o U)^{0.7} R^{0.43} \alpha^{0.54}}{E^{0.03} p^{\ 0.13}}$$

where η_o = oil viscosity in entry zone
R = effective radius
α = pressure coefficient of viscosity

Here U represents the speed, p a load parameter, and E a material parameter based on modulus and Poisson's ratio.

For ball bearings an equivalent equation is the one developed by Archard and Cowking:

$$h_{\min} = \frac{1.4(\eta_o U\alpha)^{0.74} E^{0.074}}{R^{0.74} p^{0.074}}$$

For such nonconformal systems a diagram similar to Fig. 1-3 has been suggested in which zone 2 represents elastohydrodynamic lubrication. It is difficult to think of a specific system to which the relationship exactly applies, but it may be a useful concept that the lubricant-film thickness and the friction in elastohydrodynamic lubrication bridge the gap between thick-film hydrodynamic lubrication and boundary lubrication.

A form of microelastohydrodynamic lubrication has been suggested as a mechanism for asperity lubrication under boundary conditions (see Sec. 1-5). If this suggestion is valid, the process would probably be present in the zone of mixed lubrication.

Where full-fluid-film lubrication is considered necessary but the viscosity, load, speed, and geometry are not suitable to provide full-fluid-film separation hydrodynamically, the technique of *external pressurization* can be used. Quite simply, this means feeding a fluid into a bearing at high pressure, so that the applied hydrostatic pressure is sufficient to separate the interacting surfaces of the bearing.

Externally pressurized bearings broaden the range of systems in which the benefits of full-fluid-film separation can be obtained and enable many liquids to be used successfully as lubricants which would otherwise be unsuitable. These include aqueous and other low-viscosity process fluids. Remember that the lubricant viscosity considered in Fig. 1-3 and in the various film-thickness equations is the viscosity under the relevant system conditions, especially the temperature. The viscosity of all liquids decreases with increase in temperature, and this and other factors affecting viscosity are considered in Sec. 1-4.

The viscosity and boundary lubrication properties of the lubricant completely define the lubrication performance, but many other properties are important in service. Most of these other properties are concerned with progressive deterioration of the lubricant and are described in Sec. 1-6.

1-4 LUBRICANT VISCOSITY

Viscosity of lubricants is defined in two different ways, and unfortunately both definitions are very widely used.

1-4-1 Dynamic or Absolute Viscosity

Dynamic or absolute viscosity is the ratio of the shear stress to the resultant shear rate when a fluid flows. In SI units it is measured in pascal-seconds or newton-seconds per square meter, but the centimeter-gram-second (cgs) unit, the centipoise, is more widely accepted, and

$$1 \text{ centipoise (cP)} = 10^{-3}\ \text{Pa}\cdot\text{s} = 10^{-3}\ \text{N}\cdot\text{s/m}^2$$

The centipoise is the unit of viscosity used in calculations based on the Reynolds equation and the various elastohydrodynamic lubrication equations.

1-4-2 Kinematic Viscosity

The *kinematic viscosity* is equal to the dynamic viscosity divided by the density. The SI unit is square meters per second, but the cgs unit, the centistoke, is more widely accepted, and

$$1 \text{ centistoke (cSt)} = 1\ \text{mm}^2/\text{s}$$

The centistoke is the unit most often quoted by lubricant suppliers and users.

In practice, the difference between kinematic and dynamic viscosities is not often of major importance for lubricating oils, because their densities at operating temperatures usually lie between 0.8 and 1.2. However, for some fluorinated synthetic oils with high densities, and for gases, the difference can be very significant.

The viscosities of most lubricating oils are between 10 and about 600 cSt at the operating temperature, with a median figure of about 90 cSt. Lower viscosities are more applicable for bearings than for gears, as well as where the loads are light, the speeds are high, or the system is fully enclosed. Conversely, higher viscosities are selected for gears and where the speeds are low, the loads are high, or the system is well ventilated. Some typical viscosity ranges at the operating temperatures are shown in Table 1-1.

The variation of oil viscosity with temperature will be very important in some systems, where the operating temperature either varies over a wide range or is very different from the reference temperature for which the oil viscosity is quoted.

The viscosity of any liquid decreases as the temperature increases, but the rate of decrease can vary considerably from one liquid to another. Figure 1-4 shows the change of viscosity with temperature for some typical lubricating oils. A graphical

TABLE 1-1 Typical Operating Viscosity Ranges

Lubricant	Viscosity range, cSt
Clocks and instrument oils	5–20
Motor oils	10–50
Roller bearing oils	10–300
Plain bearing oils	20–1500
Medium-speed gear oils	50–150
Hypoid gear oils	50–600
Worm gear oils	200–1000

Notes ▪ Drawings ▪ Ideas

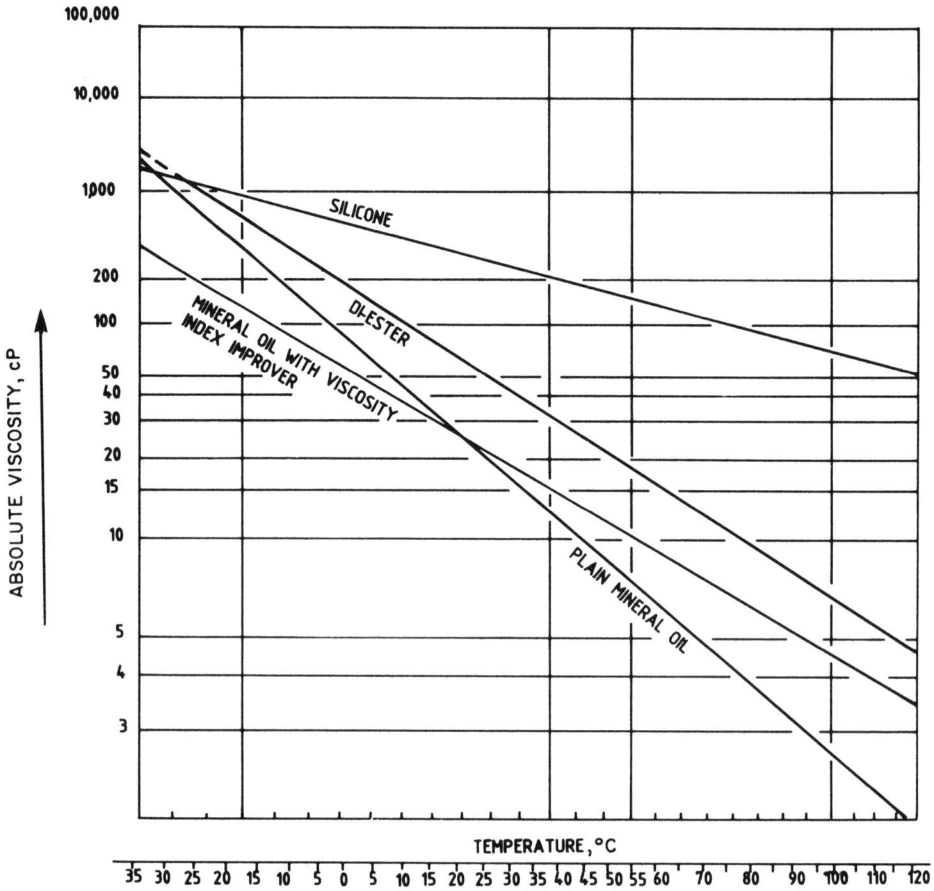

FIG. 1-4 Variation of viscosity with temperature.

presentation of this type is the most useful way to show this information, but it is much more common to quote the viscosity index (VI).

The *viscosity index* defines the viscosity-temperature relationship of an oil on an arbitrary scale in comparison with two standard oils. One of these standard oils has a viscosity index of 0, representing the most rapid change of viscosity with temperature normally found with any mineral oil. The second standard oil has a viscosity index of 100, representing the lowest change of viscosity with temperature found with a mineral oil in the absence of relevant additives.

The equation for the calculation of viscosity index of an oil sample is

$$VI = \frac{100(L - U)}{L - H}$$

wnere U = viscosity of sample in centistokes at 40°C, L = viscosity in centistokes at 40°C of oil of 0 VI having same viscosity at 100°C as test oil, and H = viscosity at 40°C of oil of 100 VI having same viscosity at 100°C as test oil.

Some synthetic oils can have viscosity indices well over 150 by the above definition, but the applicability of the definition at such high values is doubtful. The

Notes ▪ Drawings ▪ Ideas

TABLE 1-2 1977 Table of SAE Oil Ratings

SAE no.	Maximum viscosity at −18°C, cP	Viscosity at 100°C, cSt	
		Minimum	Maximum
Engine oils			
5W	1 250	3.8	
10W	2 500	4.1	
20W†	10 000	5.6	
20	.....	5.6	<9.3
30	.....	9.3	<12.5
40	.....	12.5	<16.3
50	.....	16.3	<21.9
Gear oils			
75	3 250		
80	21 600		
90	.....	14	<25
140	.....	25	<43
250	.....	43	

†15W may be used to identify 20W oils which have a maximum viscosity of 5000 cP.

viscosity index of an oil can be increased by dissolving in it a quantity (sometimes as high as 20 percent) of a suitable polymer, called a *viscosity index improver*.

The SAE viscosity rating scale is very widely used and is reproduced in Table 1-2. It is possible for an oil to satisfy more than one rating. A mineral oil of high viscosity index could meet the 20W and 30 criteria and would then be called a 20W/30 multigrade oil. More commonly a VI improved oil could meet the 20W and 50 criteria and would then be called a 20W/50 multigrade oil.

Note that the viscosity measurements used to establish SAE ratings are carried out at low shear rate. At high shear rate in a bearing, the effect of the polymer may disappear, and a 20W/50 oil at very high shear rate may behave as a thinner oil than a 20W, namely, a 15W or even 10W. In practice, this may not be important, because in a high-speed bearing the viscosity will probably still produce adequate oil-film thickness.

Theoretically viscosity index is only important where significant temperature variations apply, but in fact there is a tendency to use only high-viscosity-index oils in the manufacture of high-quality lubricant. As a result, a high viscosity index is often considered a criterion of lubricant quality, even where viscosity index as such is of little or no importance.

Before we leave the subject of lubricant viscosity, perhaps some obsolescent viscosity units should be mentioned. These are the *Saybolt viscosity* (SUS) in North America, the *Redwood viscosity* in the United Kingdom, and the *Engler viscosity* in continental Europe. All three are of little practical utility, but have been very widely used, and strenuous efforts have been made by standardizing organizations for many years to replace them entirely by kinematic viscosity.

1-5 BOUNDARY LUBRICATION

Boundary lubrication is important where there is significant solid-solid contact between sliding surfaces. To understand boundary lubrication, it is useful to first consider what happens when two metal surfaces slide against each other with no lubricant present.

In an extreme case, where the metal surfaces are not contaminated by an oxide film or any other foreign substance, there will be a tendency for the surfaces to adhere to each other. This tendency will be very strong for some pairs of metals and weaker for others. A few guidelines for common metals are as follows:

1. Identical metals in contact have a strong tendency to adhere.
2. Softer metals have a stronger tendency to adhere than harder metals.
3. Nonmetallic alloying elements tend to reduce adhesion (e.g., carbon in cast iron).
4. Iron and its alloys have a low tendency to adhere to lead, silver, tin, cadmium, and copper and a high tendency to adhere to aluminum, zinc, titanium, and nickel.

Real metal surfaces are usually contaminated, especially by films of their own oxides. Such contaminant films commonly reduce adhesion and thus reduce friction and wear. Oxide films are particularly good lubricants, except for titanium.

Thus friction and wear can usually be reduced by deliberately generating suitable contaminant films on metallic surfaces. Where no liquid lubricant is present, such a process is a type of dry or solid lubrication. Where the film-forming process takes place in a liquid lubricant, it is called boundary lubrication.

Boundary lubricating films can be produced in several ways, which differ in the severity of the film-forming process and in the effectiveness of the resulting film. The mildest film-forming process is adsorption, in which a layer one or more molecules thick is formed on a solid surface by purely physical attraction. Adsorbed films are effective in reducing friction and wear, provided that the resulting film is sufficiently thick. Figure 1-5 shows diagrammatically the way in which adsorption of a long-chain alcohol generates a thick film on a metal surface even when the film is only one molecule thick.

Mineral oils often contain small amounts of natural compounds which produce useful adsorbed films. These compounds include unsaturated hydrocarbons (olefines) and nonhydrocarbons containing oxygen, nitrogen, or sulfur atoms (known as asphaltenes). Vegetable oils and animal fats also produce strong adsorbed films and may be added in small concentrations to mineral oils for that reason. Other mild boundary additives include long-chain alcohols such as lauryl alcohol and esters such as ethyl stearate or ethyl oleate.

Adsorbed boundary films are removed fairly easily, either mechanically or by increased temperature. A more resistant film is generated by chemisorption, in which a mild reaction takes place between the metal surface and a suitable compound. Typical chemisorbed compounds include aliphatic ("fatty") acids, such as oleic and stearic acids. A chemisorbed film is shown diagrammatically in Fig. 1-6.

Even more resistant films are produced by reaction with the metal surface. The reactive compounds usually contain phosphorus, sulfur, or chlorine and ultimately produce films of metal phosphide, sulfide, or chloride on the sliding surface. These reactive additives are known as *extreme-pressure,* or EP, additives.

The processes by which modern boundary lubricant additives generate surface films may be very complex. A single additive such as trixylyl phosphate may be initially adsorbed on the metal surface, then react to form a chemisorbed film of

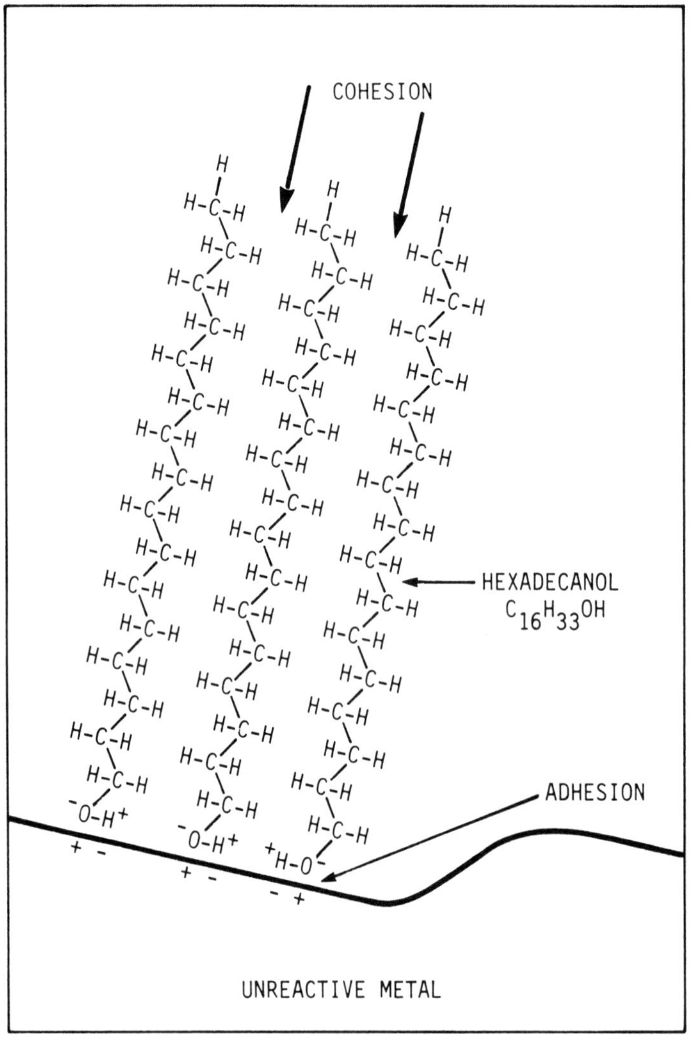

FIG. 1-5 Representation of adsorption of a long-chain alcohol. *(From Ref.* [*1-3*].)

organometallic phosphate, and finally, under severe sliding or heating, react to form metal phosphate or phosphide.

All these boundary lubricant compounds have corresponding disadvantages. As a general rule, they should be used only where the conditions of use require them. The mild, adsorbed compounds have the least undesirable side effects. They are more readily oxidized than the usual mineral-base oils and, as a result, have a higher tendency to produce corrosive acidic compounds and insoluble gums or lacquers. However, these effects are not serious, and mild antiwear additives are widely used in small quantities where sliding conditions are not severe, such as in hydraulic fluids and turbine oils.

The stronger chemisorbed additives such as fatty acids, organic phosphates, and thiophosphates are correspondingly more reactive. They are used in motor oils and gear oils. Finally, the reactive sulfurized olefines and chlorinated compounds are, in fact, controlled corrodents and are used only where the sliding conditions are very severe, such as in hypoid gearboxes and in metalworking processes.

Notes ▪ Drawings ▪ Ideas

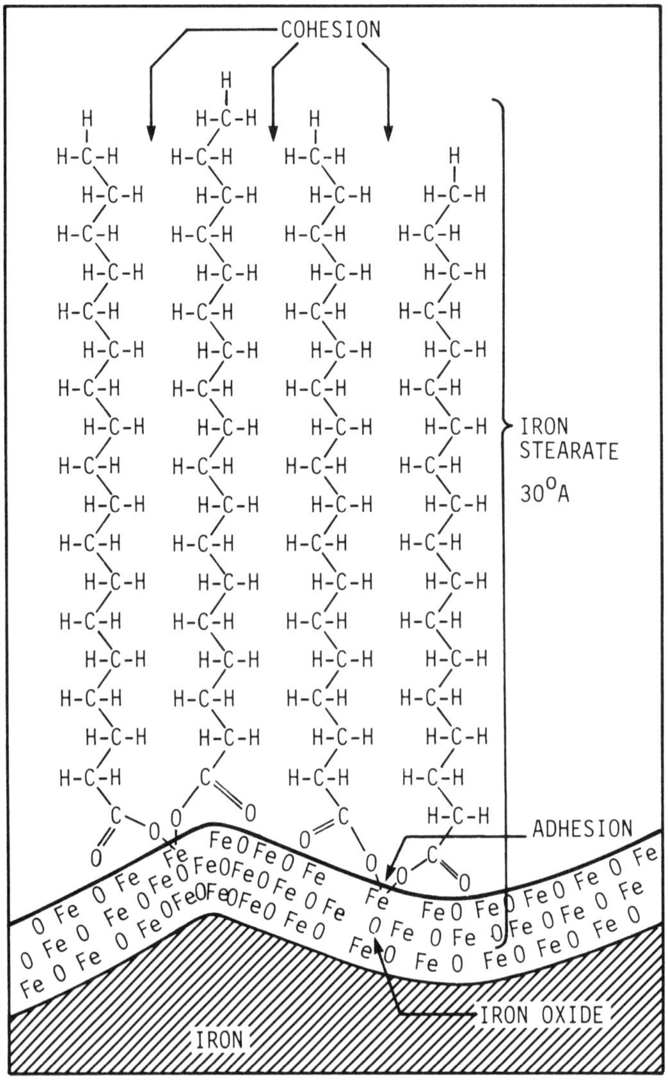

FIG. 1-6 Representation of chemisorption of a long-chain aliphatic acid. *(From Ref. [1-3].)*

Boundary lubrication is a very complex process. Apart from the direct film-forming techniques described earlier, there are several other effects which probably make an important contribution to boundary lubrication:

1. *The Rehbinder Effect* The presence of surface-active molecules adjacent to a metal surface decreases the yield stress. Since many boundary lubricants are more or less surface-active, they can be expected to reduce the stresses developed when asperities interact.

2. *Viscosity Increase Adjacent to a Metal Surface* This effect is controversial, but it seems probable that interaction between adsorbed molecules and the free ambient oil can result in a greaselike thickening or trapping of oil molecules adjacent to the surface.

3. *Microelastohydrodynamic Effects* The interaction between two asperities sliding past each other in a liquid is similar to the interaction between gear teeth, and in the same way it can be expected to generate elastohydrodynamic lubrication on a microscopic scale. The increase in viscosity of the lubricant and the elastic deformation of the asperities will both tend to reduce friction and wear. However, if the Rehbinder effect is also present, then plastic flow of the asperities is also encouraged. The term *microrheodynamic lubrication* has been used to describe this complex process.
4. *Heating* Even in well-lubricated sliding there will be transient heating effects at asperity interactions, and these will reduce the modulus and the yield stress at asperity interactions.

Boundary lubrication as a whole is not well understood, but the magnitude of its beneficial effects can be easily seen from the significant reductions in friction, wear, and seizure obtained with suitable liquid lubricants in slow metallic sliding.

1-6 DETERIORATION PROBLEMS

In theory, if the right viscosity and the right boundary properties have been selected, then the lubricated requirements will be met. In practice, there is one further complication—the oil deteriorates, and much of the technology of lubricating oils and additives is concerned with reducing or compensating for deterioration.

The three important types of deterioration are oxidation, thermal decomposition, and contamination. A fourth long-term effect is reaction with other materials in the system, which is considered in terms of compatibility. Oxidation is the most important deterioration process because over a long period, even at normal atmospheric temperature, almost all lubricants show some degree of oxidation.

Petroleum-base oils produced by mild refining techniques oxidize readily above 120°C to produce acidic compounds, sludges, and lacquers. The total oxygen uptake is not high, and this suggests that the trace compounds, such as aromatics and asphaltenes, are reacting, and that possibly in doing so some are acting as oxidation inhibitors for the paraffinic hydrocarbons present. Such mildly refined oils are not much improved by the addition of antioxidants.

More severe refining or hydrogenation produces a more highly paraffinic oil which absorbs oxygen more readily but without producing such harmful oxidation products. More important, however, the oxidation resistance of such highly refined base oils is very considerably improved by the addition of suitable oxidation inhibitors.

Most modern petroleum-base oils are highly refined in order to give consistent products with a wide operating-temperature range. Antioxidants are therefore an important part of the formulation of almost all modern mineral-oil lubricants.

The commonly used antioxidants are amines, hindered phenols, organic phosphites, and organometallic compounds. One particularly important additive is zinc diethyl dithiophosphate, which is a very effective antioxidant and also has useful boundary lubrication and corrosion-inhibition properties.

If no oxygen is present, lubricants can be used at much higher temperatures without breaking down. In other words, their thermal stability is greater than their oxidative stability. This effect can be seen for mineral oils in Table 1-3. To prevent contact of oxygen with the oil, the system must be sealed against the entry of air or purged with an inert gas such as nitrogen. Some critical hydraulic systems, such as in high-speed aircraft, are operated in this way.

Notes ▪ Drawings ▪ Ideas

TABLE 1-3 Range of Temperature Limits in Degrees Celsius for Mineral Oils as a Function of Required Life

	Life, h				
Oil condition	1	10	10^2	10^3	10^4
Thermal stability limit; insignificant oxygen present	415 to 435	385 to 405	355 to 375	320 to 340	290 to 310
Limit dependent on amount of oxygen present and presence or absence of catalysts	190 to 415	170 to 385	140 to 355	155 to 320	90 to 290
Limit imposed by oxidation where oxygen supply is unlimited; for oils containing antioxidants	175 to 190	155 to 170	125 to 140	100 to 115	80 to 90
Limit imposed by oxidation where oxygen supply is unlimited; for oils without antioxidants	155 to 165	130 to 140	95 to 110	65 to 80	35 to 50
Lower temperature limit imposed by pour point; varies with oil source, viscosity, treatment, and additives	−65 to 0	−65 to 0	−65 to 0	−65 to 0	−65 to 0

SOURCE: Ref. [1-2].

TABLE 1-4 Examples of Compatibility Problems and Possible Solutions

Problem	Solution
1. Attack by mineral oils on natural rubber	Change to nitrile rubber or neoprene
2. Attack by synthetic oils on natural rubber, nitrile, or other rubber	Change to suitable rubber for specific oil, e.g., Viton, resin-cured butyl, or EPR
3. Attack by synthetic oils on plastics or paints	Change to resistant plastics such as PTFE, polyimide, polysulfone, or polyphenylene sulfide
4. Corrosion by dissolved water	Use rust-inhibitor additives such as sulfonates
5. Corrosion by acidic degradation products	Use corrosion inhibitors such as ZDDP, or increase antioxidants to reduce degradation
6. Corrosion by additives of copper alloys or mild steel	Use less powerful EP additives, or change to corrosion-resistant metals
7. Corrosion by synthetic oils	Change to more resistant metals or platings

Notes ▪ Drawings ▪ Ideas

In high-vacuum systems such as spacecraft or electron microscopes, there is no oxygen contact. But in high vacuum an increase in temperature tends to vaporize the oil, so that high thermal stability is of little or no value. It follows that oxidative stability is usually much more important than thermal stability.

Compatibility of lubricating oils with other materials in the system is complex, and Table 1-4 lists some of the possible problems and solutions. Compatibility problems are even more complicated with synthetic lubricants and are considered further in the next section.

1-7 SELECTING THE OIL TYPE

So far most of the information in this chapter has been related to mineral oils. For almost 150 years the availability, good performance, variety, and cheapness of mineral oils have made them the first choice for most applications. They still represent over 90 percent of total lubricant use, but many other liquids are used successfully as lubricants and can provide special features which make them the best choice in particular situations.

Table 1-5 shows the most important types of lubricating oil and their advantages and disadvantages as compared with mineral oils. The natural oils comprise a wide variety of compounds of vegetable or animal origin, consisting mainly of organic esters. They all have better low-friction and boundary lubrication properties than

TABLE 1-5 Advantages and Disadvantages of Main Nonmineral Oils

Oil type	Comparison with mineral oils	
	Advantages	Disadvantages
1. Vegetable oil	Good boundary lubrication; does not cause carburization of steel in metalforming	Decomposes readily to give high viscosity or sludges and lacquers
2. Diesters, hindered esters	Higher temperature stability; high viscosity index	Some attack on rubbers and plastics
3. Polyglycol	Miscibility with water; decomposes without producing solid degradation products	Low maximum temperature
4. Silicones	High temperature stability; resistance to chemicals	Poor boundary lubrication for steel on steel
5. Phosphate ester	Fire resistance; very good boundary lubrication	Attack on rubbers and plastics; poor temperature stability
6. Chlorinated diphenyls	Fire resistance; chemical stability; boundary lubrication	Poor viscosity index; attack on plastics and copper alloys
7. Fluorocarbon	Excellent temperature and chemical stability	Price; poor viscosity index

Notes ▪ Drawings ▪ Ideas

TABLE 1-6 Some Uses of Natural Oils and Fats

Oil type	Uses
1. Rapeseed oil	*a.* To reduce friction in plain bearings where oil-film thickness is inadequate by addition of 5% to 10% to mineral oil
	b. In metal forming to give low friction and EP properties without staining or carburizing
	c. Has been used as lubricant in continuous casting
2. Castor oil	*a.* As low-viscosity hydraulic fluid for compatibility with natural rubber
	b. To give low viscous drag and good boundary lubrication in racing car engines and early aircraft engines
3. Tallow	*a.* For low friction in metal forming
4. Sperm oil	*a.* For outstanding boundary lubrication in metal cutting especially in sulfurized form; now virtually obsolete because of whale protection laws

mineral oils, but lower thermal and oxidative stability. Before mineral oils became generally available, natural oils and fats were the most common lubricants, and several are still widely used because their properties make them particularly suitable for special applications, as shown in Table 1-6.

The diesters were the first synthetic lubricating oils to be used in large quantities. Their higher thermal and oxidative stability made them more suitable than mineral oils for gas-turbine lubrication, and by about 1960 they were almost universally used for aircraft jet engines. For the even more demanding conditions of supersonic jet engines, the more complex ester lubricants such as hindered phenols and triesters were developed.

Phosphate esters and chlorinated diphenyls have very low-flammability characteristics, and this has led to their wide use where critical fire-risk situations occur,

TABLE 1-7 Some Water-Containing Lubricants

Oil type	Applications
1. Invert emulsions (water in mineral oil)	Used as hydraulic fluids for fire resistance, e.g., in coal mining. Good lubricating properties.
2. Dilute emulsions (5% mineral oil in water)	Used for fire resistance and cheapness where good lubrication properties not needed (e.g., roof jacks in coal mining).
3. "Soluble" oils (about 1% oil in water)	Used for their good cooling properties in metal cutting and grinding operations.
4. Water/Polyglycol	Used for fire resistance where increased viscosity and lack of solid degradation products are required.
5. "Synthetic" Coolants (solutions of boundary additives in water)	Used for excellent cooling and stability in metal cutting operations.

TABLE 1-8 Choice of Oil Type for Specific Properties

Property required	Choice of oil type
1. Wide range of viscosities	Mineral oil; silicone; polyglycol
2. Good boundary lubrication	Natural oil or fat; mineral oil with suitable additives; ester; phosphate ester
3. Long life	Mineral oil; silicone; fluorocarbon; ester; polyphenyl ether
4. High temperature stability	Polyphenyl ether; fluorocarbon; silicone; ester
5. Fire resistance	Emulsions; fluorocarbon; fluorosilicone; chlorinated biphenyl; phosphate ester
6. Cheapness	Emulsions; mineral oil

such as in aviation and coal mining. Their overall properties are mediocre, but are sufficiently good for use where fire resistance is particularly important.

Other synthetic fluids such as silicones, chlorinated silicones, fluorinated silicones, fluorinated hydrocarbon, and polyphenyl ethers are all used in relatively small quantities for their high-temperature stability, but all are inferior lubricants and very expensive compared with mineral oils.

Several types of water-containing fluid are used in large quantities, and these are listed in Table 1-7. They are used almost entirely to provide either fire resistance or superior cooling.

TABLE 1-9 Range of Temperature Limits in Degrees Celsius for Some Synthetic Oils as a Function of the Required Life

Name of lubricant; type of limit	Life, h				
	1	10	10^2	10^3	10^4
Polyphenyl ethers; thermal stability limit	545	520	490	455	425
Polyphenyl ethers; oxidation limit	350	330	305	280	260
Silicones; thermal stability limit	280 to 290	260 to 275	240 to 260	220 to 245	200 to 230
Esters and silicones; oxidation limit	225 to 260	215 to 245	200 to 240	185 to 220	175 to 210
Phosphate esters; thermal and oxidative limit	160	145	130	110	100
Polyphenal ethers; pour-point limit	0	0	0	0	0
Silicones and esters; pour-point limit	−60	−60	−60	−60	−60

SOURCE: Ref. [1-2].

Notes ▪ Drawings ▪ Ideas

Mineral oils can be considered as the normal, conventional oils, and alternative types are used only when they can offer some particular advantage over mineral oils. Table 1-8 summarizes the selection of oil type in relation to the special properties required.

It is difficult to give precise high-temperature limits for the use of specific oil types, because the limiting temperature depends on the required life and the amount of degradation which is acceptable. Even for water-containing lubricants, the upper temperature limit may be from 50 to 85°C depending on the required life, the degree of ventilation, and the amount of water loss which is acceptable. Table 1-9 summarizes the temperature limits for a few synthetic oils, but the limits shown should be considered only approximate.

Serious incompatibility problems can occur with lubricating oils, especially with nonmetallic materials such as rubber seals and hoses. Table 1-10 lists some satisfactory and unsatisfactory materials for use with various lubricants.

1-8 LUBRICATING GREASES

Lubricating greases are not simply very viscous oils. They consist of lubricating oils, often of quite low viscosity, which have been thickened by means of finely dispersed solids called thickeners. The effect of the thickeners is to produce a semirigid structure in which the dispersion of thickener particles is stabilized by electric charges. The liquid phase is firmly held by a combination of opposite electric charges, adsorption, and mechanical trapping. As a result, the whole grease behaves as a more or less soft solid, and there is only a very slight tendency for the oil to flow out of the grease.

Greases can probably be made from any type of lubricating oil, but in practice the majority are based on mineral oils, and only a few other base oils are of any real importance. Diesters have been used to produce greases for higher and lower tem-

TABLE 1-10 Some Compatible and Incompatible Materials for Different Oil Types

	Rubbers and plastics	
Oil type	Satisfactory	Unsatisfactory
1. Natural oils	Most rubbers, including natural rubber; most plastics	SBR rubber; highly plasticized polyethylene and polypropylene
2. Mineral oil	Nitrile rubber; neofrene; Viton; EPR; most unplasticized plastics	Natural rubber; SBR; highly plasticized plastics; polyurethanes
3. Esters	High nitrile; Viton; nylons; PPS; polyethersulfones	Natural rubber; SBR; low nitrile; polyacrylates; polyurethanes
4. Silicones	High nitrile; Viton; nylons; PPS	Natural rubber; silicone rubber; plasticized plastics
5. Phosphate ester	Resin-cured butyl rubber; EPR; PPS	Most other rubbers; many plastics

TABLE 1-11 Some Components Used in Grease Manufacture

Base oils	Thickeners	Additives
Mineral oils	Sodium soap	Antioxidants
Silicones	Lithium soap	EP additives
Diesters	Aluminum soap	Corrosion inhibitors
Chlorinated silicone	Lithium complex	Metal deactivators
Fluorocarbons	Aluminum complex	Tackiness additives
Phosphate esters	Bentonite clay	Water repellants
	PTFE	Structure modifiers
	Indanthrene dye	

peratures than mineral oils. Silicones are used for higher temperatures again, and fluorinated hydrocarbons for even higher temperatures; both these types are also used because of their chemical inertness, but the total quantities are relatively small. Phosphate esters have been used for fire resistance, and vegetable oils for compatibility with foodstuffs; but, again, the quantities are very small.

The most commonly used thickeners are soaps, which are salts of organic acids with calcium, sodium, lithium, or aluminum. The soaps take the form of fibrous particles which interlock to give a high level of stiffness at low soap concentrations. Many other substances which have been used as grease thickeners tend to be more spherical and have to be used at higher concentrations than soaps to achieve the same degree of thickening.

Most of the additives used in lubricating oils are also effective in greases. And some, such as the solid lubricants graphite and molybdenum disulfide, are much more effective in greases than in oils.

Table 1-11 lists some of the many different components which may be used in greases. The possible combinations of these components, and their different proportions, lead to an infinite range of grease formulations. In practice, a typical grease consists of a mineral oil in which are dispersed about 10 percent of a soap thickener, about 1 percent of antioxidant, and small amounts of other additives such as corrosion inhibitors, antiwear or extreme-pressure agents, and structure modifiers.

The most important physical characteristic of a grease is its relative hardness or softness, which is called *consistency*. Consistency is assessed by measuring the distance in tenths of a millimeter to which a standard metal cone penetrates the grease

TABLE 1-12 NLGI Grease Classification

NLGI number	Worked penetration at 25°C
000	445–475
00	400–430
0	355–385
1	310–340
2	265–295
3	220–250
4	175–205
5	130–160
6	85–115

Notes ▪ Drawings ▪ Ideas

under a standard load; the result is known as the *penetration.* A widely used classification of greases is that of the American National Lubricating Grease Institute (NLGI), and Table 1-12 shows the relationship between NLGI number and penetration.

The consistency of a grease varies with temperature, and there is generally an irregular increase in penetration (softening) as the temperature increases. Eventually a temperature is reached at which the grease is soft enough for a drop to fall away or flow from the bulk of the grease, and this is called the *drop point.* The drop point is usually taken to be the maximum temperature at which the grease can be used in service, but several factors confuse this situation:

1. The drop point is measured in a standard apparatus which bears no resemblance to any service equipment, so that the correlation with service use may be poor.
2. Some greases will never give a drop point because chemical decomposition begins before the thickener structure breaks down.
3. A grease may be a satisfactory lubricant above its drop point, although then it will behave like an oil rather than a grease.
4. Some greases can be heated above their drop points and will again form a grease when cooled, although normally the re-formed grease will be markedly inferior in properties.

At high temperature greases will decompose thermally or oxidatively in the same way as lubricating oils. In addition, the grease structure may break down, as explained previously, or the thickener itself may decompose. Table 1-13 depicts the general effects of temperature on lubricating greases.

A grease behaves as an extreme form of non-Newtonian fluid, and its viscous properties change when it is sheared in a feed line or a bearing. Occasionally the viscosity increases with small shear rates, but more commonly the viscosity decreases as the shear rate increases, until eventually the viscosity reaches that of the base oil. For this reason the viscosity of the base oil may be important if the grease is to be used in high-speed equipment.

The mechanism by which a grease lubricates is more complicated than for an oil, and it depends partly on the geometry of the system. Some part of the total grease fill distributes itself over the contacting surfaces and is continually sheared in the same way as an oil. This part of the grease performs the lubricating function, giving either hydrodynamic lubrication or boundary lubrication according to the load, speed, and effective viscosity.

The remainder of the grease is swept out of the path of the moving parts and remains almost completely static in the covers of a bearing or the upswept parts of a gearbox. Because of the solid nature of the grease, there is virtually no circulation or exchange between the static, nonlubricating portion and the moving, lubricating portion.

In a plain bearing or a closely fitting gearbox, a high proportion of the grease fill is being continuously sheared at the contacting surfaces. In a roller bearing or a spacious gearbox, a small proportion of the grease is continuously sheared and provides all the lubrication, while the larger proportion is inactive.

If a rolling bearing or gearbox is overfilled with grease, it may be impossible for the surplus to escape from the moving parts. Then a large quantity of grease will be continuously sheared, or "churned," and this causes a buildup of temperature which can severely damage the grease and the components. It is, therefore, important with grease lubrication to leave a void space which is sufficient to accommodate all the surplus grease, and in a ball bearing this could be more than 60 percent of the total space available.

TABLE 1-13 Temperature Limits in Degrees Celsius for Greases as a Function of Required Life

Grease; type of limit	Life, h				
	1	10	10^2	10^3	10^4
Synthetic greases; oxidation limit with unlimited oxygen present	275 to 285	255 to 265	225 to 240	200 to 225	175 to 200
Synthetic greases; drop-point limit with inorganic thickeners	250	250	250	250	250
Mineral-oil greases; upper limit imposed by drop point depends on thickener; oxidation dependent on amount of oxygen present	80 to 200	80 to 200	80 to 200	80 to 200	80 to 200
Mineral greases; oxidation limit with unlimited oxygen	185 to 200	160 to 175	135 to 150	110 to 125	85 to 100
Mineral greases; lower limit imposed by high torque	−50 to −10	−50 to −10	−50 to −10	−50 to −10	−50 to −10
Synthetic greases; lowest limit imposed by high torque	−70 to −80	−70 to −80	−70 to −80	−70 to −80	−70 to −80

SOURCE: Ref. [1-2].

The static grease which is not involved in lubrication may fulfill two useful functions. It may provide a very effective seal against the ingress of dust or other contaminants, and it can prevent loss of base oil from the grease fill. In addition, the static grease may form a reservoir to resupply the lubricated surfaces if the lubricating portion of the grease becomes depleted.

If the void space in the system is large, i.e., in a large bearing or gearbox, then usually it is desirable to use a stiffer grease to avoid the surplus grease "slumping" into the moving parts and being continuously churned. The advantages and disadvantages of grease lubrication are summarized in Table 1-14.

The selection of a grease for a specific application depends on five factors: speed, load, size, temperature range, and any grease feed system. For average conditions of speed, load, and size with no feed system, an NLGI no. 2 grease would be the normal choice, and such a grease with a mineral-oil base is sometimes known as a multipurpose grease. The effect of the various factors on selection can then be summarized in a few paragraphs.

1. *Speed* For high speeds a stiffer grease, NLGI no. 3, should be used except in plain bearings where no. 2 would usually be hard enough. For lower speeds a softer grease such as no. 1 or no. 0 should be used.
2. *Load* For high loads it may be advantageous to use EP additives or molybdenum disulfide. Because higher loads will lead to higher power consumption and therefore higher temperature, a stiffer grease such as no. 3 or a synthetic-base oil may help.
3. *Size* For large systems use a stiffer grease, no. 3 or no. 4. For very small systems use a softer grease, such as no. 1 or no. 0.
4. *Temperature Range* The drop point should be higher than the maximum predicted operating temperature. For sustained operation at higher temperatures a synthetic-base oil may be necessary. For very high temperatures, about 230°C, one of the very expensive fluorocarbon greases may be required.

TABLE 1-14 Advantages and Disadvantages of Grease Lubrication

Advantages
1. Maintain effective lubricant film on surfaces during a shutdown
2. Provide useful squeeze-film lubrication
3. Give effective sealing of rolling bearings
4. Maintain a reserve supply of lubricant in the vicinity of the bearing
5. Reduce contamination problems compared with oil
6. Provide an effective carrier for solid lubricants for antiseize or highly loaded situations
Disadvantages
1. Ineffective cooling
2. Limitations on bearing speed
3. Possible incompatibility with other similar greases
4. Lower oxidation resistance
5. Poorer storage stability

5. *Feed Systems* If the grease is to be supplied through a centralized system, usually it is desirable to use one grade softer than would otherwise be chosen (i.e., use a no. 0 instead of no. 1 or a no. 00 instead of a no. 0). Occasionally a particular grease will be found unsuitable for a centralized feed because separation occurs and the lines become plugged with thickener, but this problem is now becoming less common.

1-9 SOLID LUBRICANTS

Any solid material can act as a solid lubricant provided that it shears readily and smoothly when interposed between sliding surfaces. Some of the wide range of solids which can be used are listed in Table 1-15.

There are many other desirable properties, including the following:

1. Ability to adhere to one or both of the bearing surfaces to ensure retention in the contact area
2. Chemical stability over the required temperature range in the particular environment
3. Sufficient resistance to wear
4. Nontoxicity
5. Easy application
6. Economy

TABLE 1-15 Materials Used as Solid Lubricants

Layer-lattice compounds	
Molybdenum disulfide	Graphite
Tungsten diselenide	Tungsten disulfide
Niobium diselenide	Calcium fluoride
Graphite fluoride	
Polymers	
PTFE	PTFCE
PVF_2	FEP
Acetal	Polyimide
Polyphenylenesulfide	Polysulfones
Metals	
Silver	Gold
Tin	Lead
Barium	Gallium
Other inorganics	
Boron nitride	Molybdenum trioxide

Most of the available materials are eliminated by these requirements, and in practice almost all solid lubrication in engineering is provided by three materials—graphite, molybdenum disulfide, and polytetrafluorethylene (PTFE).

Solid lubricants can be used in several different forms, such as loose powder, adhering powder, bonded film, or solid block. In the form of a solid block, the material is often called a *dry* bearing material rather than a solid lubricant.

1-9-1 Graphite

Graphite is probably the oldest known of the three main solid lubricants, and it has ceased to be the dominant one since about 1950. It is a grayish black crystalline form of carbon in which the atoms are arranged hexagonally in monatomic layers. The strong chemical bonds between the carbon atoms give strength to the layers, so they resist bending or fracture and can carry useful loads. The bonds between the layers are relatively weak, so they slide easily over each other and can be easily separated.

When graphite is used as a lubricant, the crystals orient themselves so that the layers are parallel to the bearing surfaces. The layers then adhere fairly well to the bearing surfaces, but slide easily over each other to give low friction.

The low shearing forces, and therefore the low friction, are not an inherent property of the graphite but are strongly influenced by the presence of moisture or certain other adsorbents. If graphite is used in a very dry atmosphere, the crystal layers have quite high interlayer bonding forces, and the friction and wear are high.

The biggest advantage of graphite over molybdenum disulfide and PTFE is its electric conductivity, and it is almost universally used as a component in electric brushes. Its coefficient of friction varies from 0.05 at high loads to 0.15 at low loads, and these low values are maintained to over 500°C in air.

In block form, graphite has quite high structural integrity. It is commonly used in an impure form as graphitized carbon, in which the degree of crystallization can vary from 30 to over 80 percent of crystalline graphite. The frictional and structural properties and abrasiveness vary with the purity and degree of graphitization, and graphite technology is complex.

Graphite can be used in block form, as free powder, or as a coating deposited from dispersion in a liquid. It adheres readily to many solid surfaces, but probably its strength of adhesion is generally lower than that of molybdenum disulfide.

1-9-2 Molybdenum Disulfide

Molybdenum disulfide has also been known as a solid lubricant for centuries, but because it is similar in appearance it has often been confused with graphite. Its use has increased enormously since about 1950, and for high-technology applications it is now generally preferred to graphite. In crude form, molybdenum disulfide is found naturally, sometimes in very large quantities, as molybdenite, the most common ore of molybdenum.

Like graphite, molybdenum disulfide is a dark gray crystalline material with a hexagonal layer-lattice structure. The bond strengths within the layers are very high, while those between layers are very low. The load-carrying capacity normal to the crystal planes is therefore high, and the shear strength parallel to the crystal layers is very low.

Unlike graphite, molybdenum disulfide does not require the presence of adsorbed moisture or other vapors to give low interplanar strength. Its low friction is therefore an inherent property which is maintained in high vacuum and in dry atmospheres.

Molybdenum disulfide starts to oxidize significantly above 350°C in oxygen and 450°C in air, but the main oxidation product is molybdic oxide, which is itself a fair high-temperature lubricant. In high vacuum the disulfide is said to be stable to 1000°C, and it outgasses (evaporates) very slowly, so that it has been widely used in space.

The adhesion to metals and many other solid surfaces is excellent, and durable coatings can be produced on metal surfaces by burnishing (a coating of loose powder is rubbed into the surface to give a very thin, shiny, and strong film). The powder may be applied free or from dispersion in a volatile liquid. Durable coatings can also be obtained by sputtering, but this technique is expensive and is not widely used.

Bonded coatings are widely used, in which molybdenum disulfide powder is incorporated in almost any effective adhesive, including many polymers, natural resins, or molten solids. The performance of the softer bonded coatings is also improved if they are carefully burnished before use. The coefficient of friction of burnished films varies from 0.02 to about 0.12. But for bonded films the friction depends on the nature of the binder and the percentage composition, and it can vary from 0.02 to about 0.3.

Molybdenum disulfide is often added to oils or greases to give high load-carrying capacity, especially at low running speeds. There is also strong evidence that the addition of up to 2 percent to vehicle engine oils produces a small but significant fuel savings without any apparent disadvantages.

At one time molybdenum disulfide suffered considerable criticism, especially for reported corrosion of steels and aluminum. Some of this may have been due to its use in conjunction with graphite. Some was certainly caused by failure to understand that solid lubricants, unlike oils and greases, do not normally protect against corrosion. It is probably fair to say that molybdenum disulfide is now well understood and that, when properly used, it is a very valuable solid lubricant.

1-9-3 Polytetrafluoroethylene

Abbreviated PTFE, polytetrafluoroethylene is a polymer produced from ethylene in which all the hydrogen atoms have been replaced by fluorine atoms. This fluorination produces a material of very high chemical stability and low intermolecular bond strength, while the polymerization of an ethylene-type molecule gives long, straight molecular chains.

The result is a white solid which consists of masses of parallel long-chain molecules that slide easily past each other. This leads to the same sort of low shear strength parallel to the chains which is found in molybdenum disulfide and to a high load-carrying capacity normal to the chains, but significantly lower than that of molybdenum disulfide.

PTFE is often used in the form of solid components, occasionally in bonded coatings, and very rarely as free powder. In addition, it has been used very successfully in composites, and two types are particularly effective.

The coefficient of friction of pure PTFE varies from 0.02 at high load to about 0.1 at low load. It is rather a soft solid, so that its load-carrying capacity is limited and its wear rate is high. It therefore needs reinforcement for use in highly loaded bearings. One successful form of reinforcement is to incorporate the PTFE in the pores of a sintered metal, especially bronze. In one composite, further reinforcement is obtained by dispersing fine particles of lead in the PTFE.

A second, and probably even more successful, form of reinforcement is by means of strengthening fibers. Glass fiber or carbon fiber can be incorporated in solid PTFE components, but the resulting high structural strength is obtained at the cost of an increase in the coefficient of friction to between 0.06 and 0.2. An alternative technique is to interweave PTFE fibers and reinforcing fibers of glass, metal, rayon, or other synthetics. Some of the resulting composites have outstanding strength with low wear rate and low friction.

PTFE can be used in air to about 250°C, but in high vacuum it outgasses slowly. So it is used in spacecraft only in well-shielded locations.

Because of its high chemical stability, PTFE can be used safely in oxygen systems and in many types of chemical plants. It is nontoxic in almost all situations and is therefore used in the pharmaceutical and food industries, even in situations where low friction is not required, and as the nonstick lubricant in domestic cooking utensils.

1-9-4 Miscellaneous Solid Lubricants

Other solid lubricants are used to a relatively minor degree, where they have specific advantages. They can be classified in three broad categories: inorganics, polymers, and metals.

The inorganics include a number of materials similar to molybdenum disulfide, known generally as the lubricating dichalcogenides. None of these occurs naturally, and the synthetic materials are relatively expensive. Tungsten disulfide has a higher oxidation temperature, and both tungsten disulfide and tungsten diselenide oxidize more slowly than molybdenum disulfide. Niobium diselenide has better electric conductivity and has been used in electric contact brushes, but in fact molybdenum disulfide composites have been shown to be equally satisfactory.

Other inorganics have been used for their much higher temperature limits, and these include molybdic oxide, boron nitride, graphite fluoride, and calcium fluoride.

The low friction and chemical inertness of PTFE make it difficult to bond to other materials, and two other fluorinated polymers have been recommended for their better bonding behavior: polyvinylfluoride (PVF_2) and polytrifluorochloroethylene (PTFCE). But in both cases the advantages of better bonding and slightly higher structural strength are offset by higher friction.

For higher temperatures, polyimide, polysulfones, and polyphenylene sulfide can be used unlubricated. Other polymers such as nylons, acetals, and phenolics are occasionally used unlubricated where sliding speeds are low, but they require lubrication by oil, grease, or water for really useful performance.

Silver, gold, and tin have useful antigalling properties in slow sliding, but metallic coatings are mainly used as lubricants only in high vacuum, where silver, gold, barium, gallium, and lead have all been used successfully.

1-10 GAS LUBRICATION

Gases can be used to provide full-fluid-film lubrication in the same ways as liquids in hydrodynamic and externally pressurized bearings. The physical laws governing the behavior are the same for both liquids and gases, but the very low viscosities of gases lead to considerable practical differences in their use, especially in self-pressurizing, or "gas-dynamic," bearings:

1. Operating speeds are much higher to compensate for low viscosity.
2. Surface finish and precision must be better because of the much smaller lubricant-film thickness.

3. Lubricant flow rate is higher for the same pressure differential.
4. Load-carrying capacity is generally low.

As a result, gas bearings tend to be small, high-speed, and lightly loaded, with tight tolerances and high-quality surface finishes. The overall design and manufacturing cost is high, and they are mainly used in high-technology applications.

Any gas or vapor can be used provided that it is chemically stable under the operating conditions and does not attack any of the system materials. If no chemical change takes place, there is no upper temperature limit to the use of a gas, and the viscosity increases as temperature increases.

Air is the most common gas used in gas lubrication. Nitrogen or helium may be used where inertness is important. Otherwise, any gas which is available can be used, especially if it is available at high enough pressure for external pressurization.

Some of the advantages of gas lubrication are high precision, very low friction, cleanliness, and ready availability of lubricant. The greatest potential advantage is the wide temperature range. In theory, it should be possible to design a gas bearing to operate from −250 to +2000°C. The corresponding disadvantages include the demanding design and construction requirements, the low load-carrying capacity, and the need for a very clean gas supply. Examples of important applications of gas bearings are dentists' air-turbine drills, high-precision grinding spindles, and inertial gyroscopes.

1-11 LUBRICANT FEED SYSTEMS

In many lubricated components, no feed system is needed, because the initial lubricant fill is sufficient to last the required life. A feed system becomes necessary when the lubricant must be replaced or replenished, for one of the following reasons:

1. Temperature is too high, so that the lubricant must be removed and replaced by a fresh charge of cooler lubricant.
2. Lubricant becomes depleted by leakage or creepage and must be topped up.
3. Lubricant decomposes and must be replaced with fresh charge.
4. Lubricant becomes contaminated and must be replaced with clean material.

Where the rate of loss or deterioration is relatively low, it will be sufficient to provide facility for occasional topping up by means of an oil can or a grease gun, provided that access to the lubricated component is available. Where this occasional manual topping up is not adequate, a lubricant feed system will be needed. It is beyond the scope of this chapter to describe the whole range and design of lubricant feed systems available. It is only possible to give a brief description of the main types and the factors involved in selecting them.

1-11-1 Internal Circulation

One obvious way to reduce oil temperature, slow down the increase in contamination, and increase the life is simply to increase the quantity of oil supplied in the initial fill. This requires an increase in the volume of space available for oil or, in other words, the creation of an oil reservoir or sump adjacent to the lubricated bearings or gears.

Circulation of the oil can be ensured by arranging for the moving parts to dip below the surface of the oil. But they should not be completely submerged because the resulting viscous drag and churning of the oil lead to excessive power consumption and heating. For slow-moving components this problem is not serious, but for high speeds the depth of immersion is critical, and the following guidelines are useful:

1. Gears should be immersed to twice the tooth height. In a vertical train, the oil level should be just below the shaft of the lowest gear.
2. Rolling bearings should be immersed to halfway up to the lowest rollers or balls.
3. Crankshafts should be immersed so that the oil level is just above the big-end bearings at their lowest point.
4. For slow operation the oil level should be higher than for higher-speed systems.

Oil is carried by the partly submerged components to contacting surfaces and is also spread by splashing. Where transfer by these two mechanisms is inadequate, the oil feed can be improved by the use of rings or disks, as shown in Fig. 1-7. Both operate by providing a larger surface with higher peripheral speed to transfer the oil; but they do not cause excessive viscous drag because they are both uniform in shape.

Disks have an advantage over rings in that they can be designed to propel oil axially as well as radially, and this is particularly useful for bearing lubrication. Usually plain bearings cannot be adequately lubricated by partial immersion in oil unless the oil flow is augmented by a ring or a disk. If a weir is incorporated, part of the splashed oil can be trapped and directed to critical locations.

1-11-2 Topping Up Systems

Where the main problem is loss of oil by leakage or creep, it may be sufficient to set up a wick or drip feed to provide a small-scale supply. Wicks or pads consist of porous or permeable materials such as felt which transfer oil by capillary action to the bearing surfaces. The pads may form a path from a reservoir to the bearing or may simply contain a small initial oil fill to increase the quantity available and feed it slowly to the bearing. This latter approach is commonly used in small electric motors.

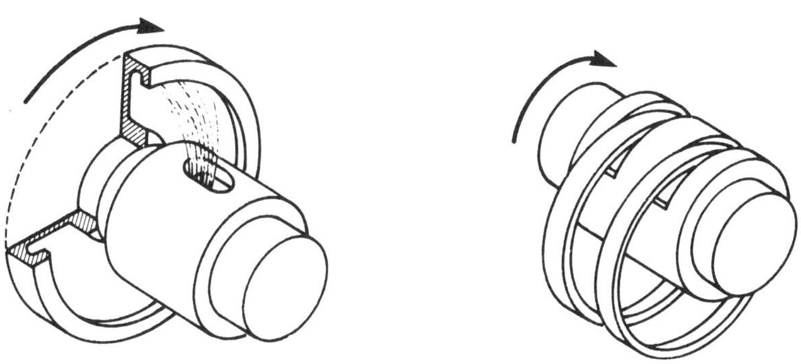

(a) (b)

FIG. 1-7 Ring and disk lubrication. (*a*) Disk; (*b*) ring. *(From Ref.* [*1-1*].*)*

Drip feeds consist of small reservoirs mounted above the bearing or gears and equipped with a feed tube with some form of flow regulation and usually a sight glass. They can be used to provide a much higher flow rate than a wick.

1-11-3 Centralized Total-Loss Systems

Wick and drip feeds are examples of total-loss systems, in which no attempt is made to collect the oil after feeding it to the bearings or gears. Far more sophisticated total-loss systems can be used to supply oil or grease to a number of separate components.

The requiremens of centralized total-loss sytems are basically very simple. A typical system consists of a reservoir, a single pump to pressurize a manifold, and a number of controllers, each regulating the feed to a single lubrication point. Alternatively the flow to several lines may be controlled by a multipiston pump, in which individual single-cylinder piston pumps are operated by cams on a common camshaft immersed in the lubricant. The number of outlets can vary from one to several hundred.

The main advantage of centralized total-loss systems is that they reduce the labor required where a large number of components need relubricating. They are also valuable where the lubrication points are not readily accessible. Their disadvantages are that they do not provide any form of cooling or removal of contaminants, and there is no recovery of used lubricant.

1-11-4 Oil Mist or Fog Systems

One type of total-loss system which has become widely used in recent years is oil mist or fog, in which fine droplets of oil are carried by a stream of air from a reservoir to a bearing or gears. The mist or fog of oil in air is produced by passing the airstream through the reservoir at low speed and low pressure. The oil is usually formulated to have low surface tension. The resulting dispersion of oil droplets is passed through steel, copper, or plastic pipes to the vicinity of the lubricated component. Passage through fine nozzles, or reclassifiers, then increases the linear flow rate to something over 45 meters per second (m/s) and thus causes the oil droplets to coalesce, producing a flow of liquid oil to the lubrication point.

Oil mist and fog systems have two advantages over other total-loss systems in that the oil supply rate can be very low, resulting in a clean system, while the airflow gives a significant amount of cooling.

1-11-5 Oil-Circulation Systems

The most sophisticated centralized systems are those in which the lubricant is collected after use and returned to the reservoir for recirculation. The basic requirements of such a system are a reservoir, a pump, possibly a flow divider or proportioner, feed lines, and return lines.

In practice a full-circulation system is likely to be more complex and to include many of or all the following components:

Multiple or divided reservoirs

Heaters

Coolers

Oil-level warning devices

Full-flow filtration to protect the pump
Bypass filtration
Pressure switches and alarms
Water separators
Chip detectors
Sampling points
Sight glass

Circulation systems give a high degree of control over the quality and quantity of oil supplied to each component, enabling the cleanliness and the temperature of the oil to be controlled. Their only disadvantage is their complexity and therefore their cost.

1-12 LUBRICANT STORAGE

The storage of lubricants, like any other class of goods, depends first on making rational decisions about the number of varieties and the quantities which it is necessary to store. The special factors which then need to be considered in storing lubricants are as follows:

1. Lubricants are an integral part of the precision components in which they are used. They must therefore be treated as precision components and protected carefully against contamination by dirt, water, or other materials. Never store them in the open.
2. Because most lubricants are liquids, they have no characteristic shapes, and it is very easy to use the wrong lubricant in a machine. This is always undesirable and sometimes catastrophic. So it is important to label lubricant containers carefully and to control their issue and use.
3. Some lubricants deteriorate in storage, and it is important to use supplies in proper rotation and ensure that storage lives are not exceeded.
4. Many lubricants are flammable, and special precautions are necessary to reduce fire risk.
5. Lubricants are slippery, and spillages can cause accidents. Floor gratings and drainage channels should be supplied, and absorbent powders or granules kept available to absorb spilled oil.
6. Oil drums are very convenient and satisfactory containers for storing lubricants, but water can collect in the recessed top and enter even through sealed apertures to contaminate the contents. Drums should always be stored horizontally and never upright.

REFERENCES

1-1 A. R. Lansdown, *Lubrication: A Practical Guide to Lubricant Selection,* Pergamon, New York, 1982.

1-2 M. J. Neale (ed.), *The Tribology Handbook,* Butterworth, London, 1973.

1-3 D. Godfrey, ''Boundary Lubrication,'' in P. M. Ku (ed.), *Interdisciplinary Approach to Friction and Wear,* NASA SP-181, 1968.

chapter 2
GASKETS AND SEALS

DANIEL E. CZERNIK
Director of Product Engineering
Fel-Pro Inc.
Skokie, Illinois

R. BRUCE HOPKINS, Ph.D., P.E.
The Hopkins Engineering Co., P.C.
Cedar Falls, Iowa

In the field of gaskets and seals, the former are generally associated with sealing mating flanges while the latter are generally associated with sealing reciprocating shafts or moving parts. Some designers refer to gaskets as static seals and consider seals to be dynamic sealing components. This chapter is divided in this way. Part 1 covers gaskets, and Part 2 discussed seals.

PART 1 GASKETS

DANIEL E. CZERNIK

2-1 DEFINITION

A *gasket* is a material or combination of materials clamped between two separable members of a mechanical joint. Its function is to effect a seal between the members (flanges) and maintain the seal for a prolonged period. The gasket must be capable of sealing mating surfaces, must be impervious and resistant to the medium being

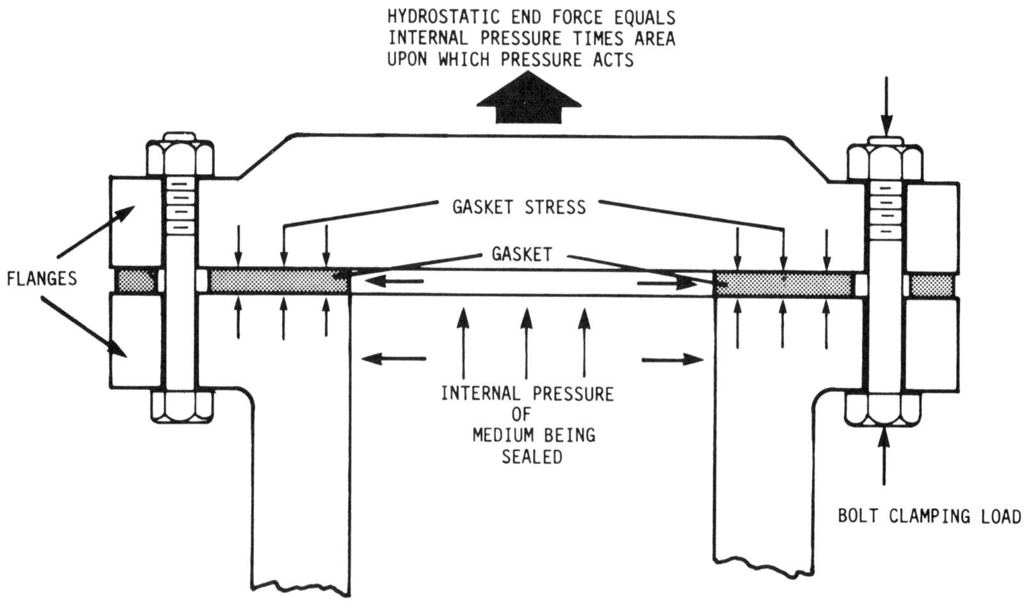

FIG. 2-1 Nomenclature of a gasketed joint.

sealed, and must be able to withstand the application temperature. Figure 2-1 depicts the nomenclature associated with a gasketed joint.

2-2 STANDARD CLASSIFICATION SYSTSEM FOR NONMETALLIC GASKET MATERIALS†

This classification system provides a means for specifying or describing pertinent properties of commercial nonmetallic gasket materials. Materials composed of asbestos, cork, cellulose, and other organic or inorganic materials in combination with various binders or impregnants are included. Materials normally classified as rubber compounds are not included, since they are covered in ASTM Method D 2000 (SAE J200). Gasket coatings are not covered since details are intended to be given on engineering drawings or in separate specifications.

This classification is based on the principle that nonmetallic gasket materials can be described in terms of specific physical and mechanical characteristics. Thus, users of gasket materials can, by selecting different combinations of statements, specify different combinations of properties desired in various parts. Suppliers, likewise, can report properties available in their products.

In specifying or describing gasket materials, each *line call-out* shall include the number of this system (minus date symbol) followed by the letter F and six numerals, for example, ASTM F104 (F125400). Since each numeral of the call-out represents a characteristic (as shown in Table 2-1), six numerals are always required. The numeral 0 is used when the description of any characteristic is not desired. The numeral 9 is used when the discription of any characteristic (or related test) is specified by some supplement to this classification system, such as notes on engineering drawings.

†Ref. [2-1] (ANSI/ASTM F104).

TABLE 2-1 Basic Physical and Mechanical Characteristics

Basic six-digit number	Basic characteristic
First numeral	Type of material (the principal fibrous or particulate reinforcement material from which the gasket is made) shall conform to the first numeral of the basic six-digit number as follows: 0 = not specified 1 = asbestos or other inorganic fibers (type 1) 2 = cork (type 2) 3 = cellulose or other organic fibers (type 3) 4 = fluorocarbon polymer 9 = as specified†
Second numeral	Class of material (method of manufacture or common trade designation) shall conform to the second numeral of the basic six-digit number as follows: When first numeral is 1, for second numeral 0 = not specified 1 = compressed asbestos (class 1) 2 = beater addition asbestos (class 2) 3 = asbestos paper and millboard (class 3) 9 = as specified† When first numeral is 2, for second numeral 0 = not specified 1 = cork composition (class 1) 2 = cork and elastomeric (class 2) 3 = cork and cellular rubber (class 3) 9 = as specified† When first numeral is 3, for second numeral 0 = not specified 1 = untreated fiber—tag, chipboard, vulcanized fiber, etc. (class 1) 2 = protein treated (class 2) 3 = elastomeric treated (class 3) 4 = thermosetting resin treated (class 4) 9 = as specified† When first numeral is 4, for second numeral 0 = not specified 1 = sheet PTFE 2 = PTFE of expanded structure 3 = PTFE filaments, braided or woven 4 = PTFE felts 5 = filled PTFE 9 = as specified†
Third numeral	Compressibility characteristics, determined in accordance with 8.2, shall conform to the percentage indicated by the third numeral of the basic six-digit number (example: 4 = 15 to 25%): 0 = not specified 5 = 20 to 30% 1 = 0 to 10% 6 = 25 to 40% 2 = 5 to 15%‡ 7 = 30 to 50% 3 = 10 to 20% 8 = 40 to 60% 4 = 15 to 25% 9 = as specified†

TABLE 2-1 Basic Physical and Mechanical Characteristics (*Continued*)

Fourth numeral	Thickness increase when immersed in ASTM no. 3 oil, determined in accordance with 8.3, shall conform to the percentage indicated by the fourth numeral of the basic six-digit number (example: 4 = 15 to 30%): 0 = not specified 5 = 20 to 40% 1 = 0 to 15% 6 = 30 to 50 % 2 = 5 to 20% 7 = 40 to 60% 3 = 10 to 25% 8 = 50 to 70% 4 = 15 to 30% 9 = a specified†
Fifth numeral	Weight increase when immersed in ASTM no. 3 oil, determined in accordance with 8.3, shall conform to the percentage indicated by the fifth numeral of the basic six-digit number (example: 4 = 30% maximum): 0 = not specified 5 = 40% max. 1 = 10% max. 6 = 60% max. 2 = 15% max. 7 = 80% max. 3 = 20% max. 8 = 100% max. 4 = 30% max. 9 = as specified†
Sixth numeral	Weight increase when immersed in water, determined in accordance with 8.3, shall conform to the percentage indicated by the sixth numeral of the basic six-digit number (example: 4 = 30% maximum): 0 = not specified 5 = 40% max. 1 = 10% max. 6 = 60% max. 2 = 15% max. 7 = 80% max. 3 = 20% max. 8 = 100% max. 4 = 30% max. 9 = as specified†

†On engineering drawings or other supplement to this classification system. Suppliers of gasket materials should be contacted to find out what line call-out materials are available. Refer to ANSI/ASTM F104 for further details (Ref. [2-1]).

‡From 7 to 17% for type 1, class 1 compressed asbestos sheet.

2-3 GASKET PROPERTIES, TEST METHODS, AND THEIR SIGNIFICANCE IN GASKETED JOINTS

Table 2-2 lists some of the most significant gasket properties which are associated with creating and maintaining a seal. This table also shows the test method as well as the significance of each property in a gasket application.

2-4 PERMEABILITY PROPERTIES

The primary criterion for a material to be impervious to a fluid is to achieve a sufficient density to eliminate voids which might allow capillary flow of the fluid through the construction. This requirement may be met in two ways: by compressing the material to fill the voids and/or partially or completely filling them during fabrication by means of binders and fillers. Also, to maintain its impermeability for a prolonged time, the constituents of the material must be able to resist degradation

Notes ▪ Drawings ▪ Ideas

TABLE 2-2 Identification, Test Method, and Significance of Various Properties Associated with Gasket Materials

Property	Test method	Significance in gasket applications
Sealability	Fixtures per ASTM F37-62T	Resistance to fluid passage
Heat resistance	Exposure testing at elevated temperatures	Resistance to thermal degradation
Oil and water immersion characteristics	ASTM D-1170	Resistance to fluid attack
Antistick characteristics	Fixture testing at elevated termperatures	Ability to release from flanges after use
Stress vs. compression and spring rates	Various compression test machines	Sealing pressure at various compressions
Compressibility and recovery	ASTM F36-61T	Ability to follow deformation and deflection; indentation characteristics
Creep relaxation and compression set	ASTM F38-62T and D-395-59	Related to torque loss and subsequent loss of sealing pressure
Crush and extrusion characteristics	Compression test machines	Resistance to high loadings and extrusion characteristics at room and elevated temperatures

and disintegration resulting from chemical attack and temperature of the application [2-2].

Most gasket materials are composed of a fibrous or granular base material, forming a basic matrix or foundation, which is held together or strengthened with a binder. The choice of combinations of binder and base material depends on the compatibility of the components, conditions of the sealing environment, and load-bearing properties required for the application.

Some of the major constituents and the properties which are related to impermeability are listed here.

2-4-1 Base Materials—Nonmetallic

CORK AND CORK-RUBBER. High compressibility allows easy density increase of the material, thus enabling an effective seal at low flange pressures. The temperature limit is approximately 250°F (121°C) for cork and 300°F (149°C) for cork-rubber compositions. Chemical resistance to water, oil, and solvents is good, but resistance to inorganic acids, alkalies, and oxidizing environments is poor. It conforms well to distorted flanges.

CELLULOSE FIBER. Cellulose has good chemical resistance to most fluids except strong acids and bases. The temperature limitation is approximately 300°F (149°C). Changes in humidity may result in dimensional changes and/or hardening.

ASBESTOS FIBER. This material has good heat resistance to 800°F (427°C) and is noncombustible. It is almost chemically inert (crocidolite fibers, commonly known as blue asbestos, resist even inorganic acids) and has very low compressibility. Binder dictates resistance to temperature and medium to be sealed.

NONASBESTOS FIBERS. A number of nonasbestos fibers are being used in gaskets. Some of these are glass, carbon, aramid, and ceramic. These fibers are expensive and are normally used only in small amounts. Temperature limits from 750 to 2400°F (399 to 1316°C) are obtainable. Use of these fillers is an emerging field today, and suppliers should be contacted before these fibers are specified for use.

2-4-2 Binders and Fillers

RUBBER. Rubber binders provide varying temperature and chemical resistance depending on the type of rubber used. These rubber and rubberlike materials are used as binders and, in some cases, gaskets:

1. *Natural* This rubber has good mechanical properties and is impervious to water and air. It has uncontrolled swell in petroleum oil and fuel and chlorinated solvents. The temperature limit is 250°F (121°C).
2. *Styrene/Butadiene* This rubber is similar to natural rubber but has slightly improved properties. The temperature limit also is 250°F (121°C).
3. *Butyl* This rubber has excellent resistance to air and water, fair resistance to dilute acids, and poor resistance to oils and solvents. It has a temperature limit of 300°F (149°C).
4. *Nitrile* This rubber has excellent resistance to oils and dilute acids. It has good compression set characteristics and has a temperature limit of 300°F (149°C).
5. *Neoprene* This rubber has good resistance to water, alkalies, nonaromatic oils, and solvents. Its temperature limit is 250°F (121°C).
6. *Ethylene Propylene Rubber* This rubber has excellent resistance to hot air, water, coolants, and most dilute acids and bases. It swells in petroleum fuels and oils without severe degradation. The temperature limit is 300°F (149°C).
7. *Acrylic* This rubber has excellent resistance to oxidation, heat, and oils. It has poor resistance to low temperature, alkalies, and water. The temperature limit is 450°F (232°C).
8. *Silicone* This rubber has good heat stability and low-temperature flexibility. It is not suitable for high mechanical pressure. Its temperature limit is 600°F (316°C).
9. *Viton* This rubber has good resistance to oils, fuel, and chlorinated solvents. It also has excellent low-temperature properties. Its temperature limit is 600°F (316°C).
10. *Fluorocarbon* This rubber has excellent resistance to most fluids, except synthetic lubricants. The temperature limit is 500°F (260°C).

RESINS. These usually possess better chemical resistance than rubber. Temperature limitations depend on whether the resin is thermosetting or thermoplastic.

TANNED GLUE AND GLYCERINE. This combination produces a continuous gel structure throughout the material, allowing sealing at low flange loading. It has good chemical resistance to most oils, fuels, and solvents. It swells in water but is not soluble. The temperature limit is 200°F (93°C). It is used as a saturant in cellulose paper.

FILLERS. In some cases, inert fillers are added to the material composition to aid in filling voids. Some examples are barytes, asbestine, and cork dust.

Notes ▪ Drawings ▪ Ideas

2-4-3 Reinforcements

Some of the properties of nonmetallic gasket materials can be improved if the gaskets are reinforced with metal or fabric cores. Major improvements in torque retention and blowout resistance are normally seen. Traditionally, perforated or upset metal cores have been used to support gasket facings. A number of designs have been utilized for production. Size of the perforations and their frequency in a given area are the usual specified parameters.

Adhesives have been developed that permit the use of an unbroken metal core to render support to a gasket facing. Laminated composites of this type have certain characteristics that are desired in particular gaskets [2-3].

2-4-4 Metallic Materials

ALUMINUM. This metal has good conformability and thermal conductivity. Depending on the alloy, aluminum suffers tensile strength loss as a function of temperature. Normally it is recommended up to 800°F (427°C). It is attacked by strong acids and alkalies.

COPPER. This metal has good corrosion resistance and heat conductivity. It has ductility and excellent flange conformability. Normally 900°F (482°C) is considered the upper service temperature limit.

STEEL. A wide variety of steels are used in gasketing. Mild steel to stainless steel has been utilized. A high clamping load is required. Temperature limits range from 1000 to 2100°F (538 to 1149°C) depending on the alloy.

2-5 LOAD-BEARING PROPERTIES

2-5-1 Conformability and Pressure

Since sealing conditions vary widely depending on the application, it is necessary to vary the load-bearing properties of the gasket elements in accordance with these conditions. Figure 2-2 illustrates stress-compression curves for several gasket components and indicates the difference in the stress-compression properties used for different sealing locations.

Gasket thickness and compressibility must be matched to the rigidity, roughness, and unevenness of the mating flanges. An entire seal can be achieved only if the stress level imposed on the gasket at clampup is adequate for the specific material. Minimum seating stresses for various gasket materials are listed later in this chapter. In addition, the load remaining on the gasket during operation must remain high enough to prevent blowout of the gasket. During operation, the hydrostatic end force, which is associated with the internal pressure, tends to unload the gasket. Figure 2-3 is a graphical representation of a gasketed joint depicting the effect of the hydrostatic end force [2-4].

The bolt should be capable of handling the maximum load imposed on it without yielding. The gasket should be capable of sealing at the minimum load resulting on it and should resist blowout at this load level.

Gaskets fabricated from compressible materials should be as thin as possible [2-

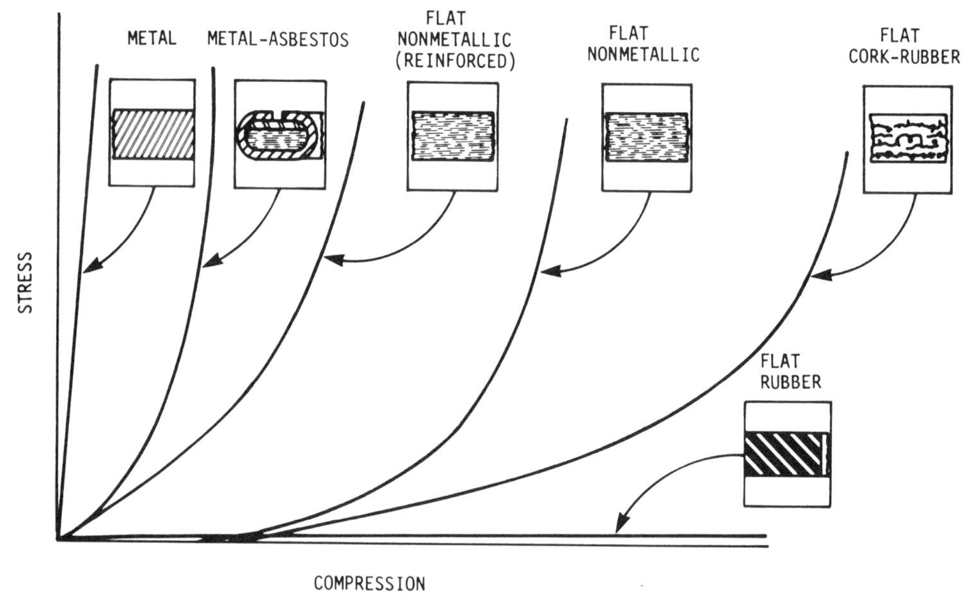

FIG. 2-2 Stress versus compression for various gasket materials.

5]. The thickness should be no greater than that which is necessary for the gasket to conform to the unevenness of the mating flanges. The unevenness is associated with surface finish, flange flatness, and flange warpage during use. It is important to use the gasket's unload curve in considering its ability to conform. Figure 2-4 depicts typical load-compression and unload curves for nonmetallic gaskets.

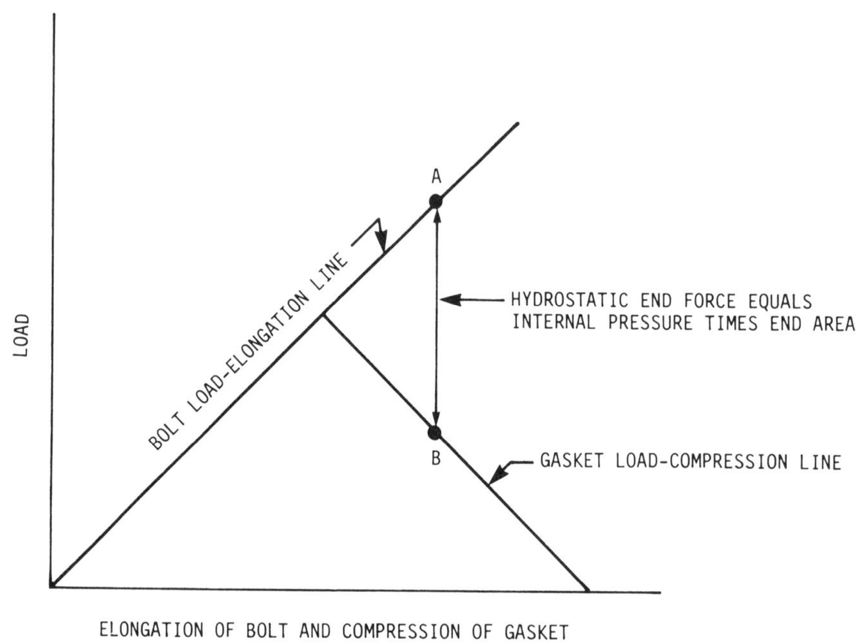

FIG. 2-3 Graphical representation of a gasketed joint and effect of hydrostatic end force. *A*, Maximum load on gasket; *B*, minimum load on gasket.

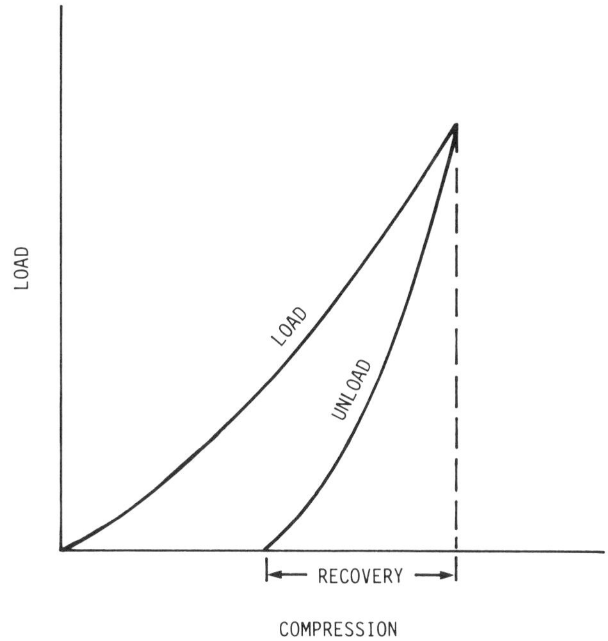

FIG. 2-4 Load-compression and unload curve for a typical nonmetallic gasket material.

The unload curve determines the recovery characteristics of the gasket which are required for conformance. Metallic gaskets will show no change in their load and unload curves unless yielding occurs. Load-compression curves are available from gasket suppliers.

Some advantages of thin gaskets over thick gaskets are

1. Reduced creep relaxation and subsequent torque loss
2. Less distortion of mating flanges
3. Higher resistance to blowout
4. Fewer voids through which sealing media can enter, so less permeability
5. Lower thickness tolerances
6. Better heat transfer

A common statement in the gasket industry is, "Make the gasket as thin as possible and as thick as necessary."

The following paragraphs describe some of the gasket's design specifications which need to be considered for various applications. A large array of gasket designs and sealing applications are used, and more are coming into use daily. Gaskets are constantly being improved for higher and higher performance.

In high-pressure, clamp load, and temperature applications, a high-spring-rate (stress per unit compression) material is necessary to achieve high loading at low compression in order to seal the high pressures developed. These applications generally rely on sealing resulting from localized yielding under the unit loading. In addition to the high spring rate, high heat resistance is mandatory. To economically satisfy these conditions, metal is the most commonly used material.

In applications where close tolerances in machining (surface finish and parallelism) are obtainable, a solid steel construction may be used. In those situations where close machining and assembly is not economical, it is necessary to sacrifice some gasket rigidity to allow for conformability. In such cases, conformability exceeding that resulting from localized yielding must be inherent in the design. The metal can be corrugated, or a composite design consisting of asbestos could be used to gain the conformability required.

In very-high-pressure applications, flat gaskets may not have adequate recovery to seal as the hydrostatic end force unseats the gaskets [2-6]. In these cases, various types of self-energized metal seals are available. These seals utilize the internal pressure to achieve high-pressure sealing. They require careful machining of the flanges and have some fatigue restrictions.

In applications where increased surface conformity is necessary and lower temperatures are encountered, asbestos and/or other nonmetallic materials can be used under the limitations noted earlier.

Elastomeric inserts are used in some fluid passages where conformity with sealing surfaces and permeability are major problems and high fluid pressures are encountered. Since the inserts have low spring rates, they must be designed to have appropriate contact areas and restraint in order to effect high unit sealing stresses for withstanding the internal pressures. The inserts also have high degrees of recovery, which allow them to follow high thermal distortions normally associated in the mating flanges. Compression set and heat-aging characteristics must also be considered when elastomeric inserts are used.

2-5-2 Creep and Relaxation

After the initial sealing stress is applied to a gasket, it is necessary to maintain a sufficient sealing stress for the designed life of the unit or equipment. All materials exhibit, in varying degrees, a decrease in applied stress as a function of time, commonly referred to as *stress relaxation.* The reduction of stress on a gasket is actually a combination of two major factors: stress relaxation and creep (compression drift). By definition,

> *Stress relaxation* is a reduction in stress on a specimen under constant strain ($d\sigma/dt$; e = constant).
>
> *Creep* (compression drift) is a change in strain of a specimen under constant stress (de/dt; σ = constant).

In a gasketed joint, stress is applied by tension in a bolt or stud and transmitted as a compressive force to the gasket. After loading, stress relaxation and creep occur in the gasket, causing corresponding lower strain and tension in the bolt. This process continues indefinitely as a function of time. The change in tension of a bolt is related to the often quoted "torque loss" associated with a gasket application. Since the change in stress is due to two primary factors, a more accurate description of the phenomenon would be *creep relaxation,* from now on called *relaxation.*

Bolt elongation, or stretch, is linearly proportional to bolt length. The longer the bolt, the higher the elongation. The higher the elongation, the lower the percentage loss for a given relaxation. Therefore, the bolts should be made as long as possible for best torque retention.

Relaxation in a gasket material may be measured by applying a load on a specimen by means of a strain-gauged bolt-nut-platen arrangement as standardized by ASTM F38-62T. Selection of materials with good relaxation properties will result in the highest retained torque for the application. This results in the highest remaining stress on the gasket which is desirable for long-term sealing.

Notes ▪ Drawings ▪ Ideas

The amount of relaxation increases as thickness is increased for a given gasket material. This is another reason why the thinnest gasket that will work should be selected. Figure 2-5 depicts the relaxation characteristics as a function of thickness for a particular gasket design.

Note that as clamping stress is increased, relaxation is decreased. This is the result of more voids being eliminated as the stress level is increased.

2-5-3 Effect of Geometry

The gasket's shape factor has an important effect on its relaxation characteristics. This is particularly true in the case of soft packing materials.

Much of the relaxation of a material may be attributed to the releasing of forces through lateral expansion. Therefore, the greater the area available for lateral expansion, the greater the relaxation. The *shape factor* of a gasket is the ratio of the area of one load face to the area free to bulge. For circular or annular samples, this may be expressed as

$$\text{Shape factor} = \frac{1}{4t}(\text{OD} - \text{ID}) \tag{2-1}$$

where t = thickness of gasket
OD = outside diameter
ID = inside diameter

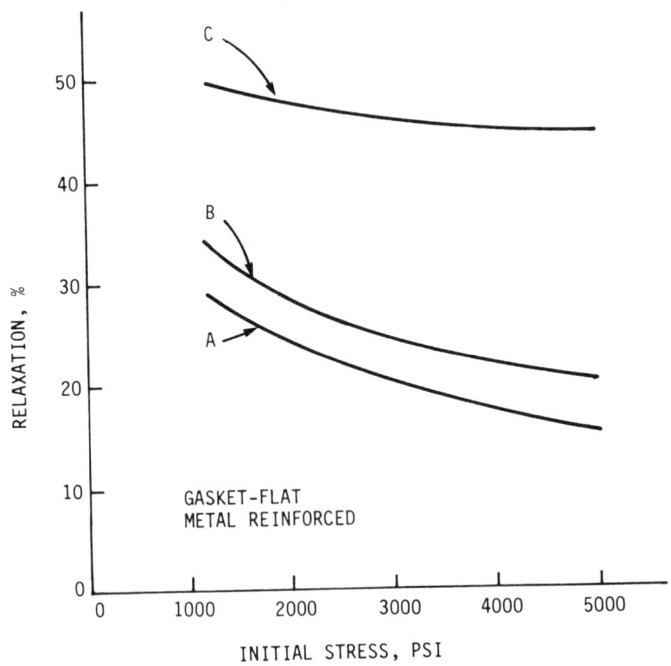

FIG. 2-5 Relaxation versus stress on a gasket A, 0.030 in/0.035 in thick; B, 0.042 in/0.047 in thick; C, 0.062 in/0.065 in thick.

As the area to bulge increases, the shape factor decreases, and the relaxation will increase as the retained stress decreases. Figure 2-6 depicts the effect of shape factor on the gasket's ability to retain stress.

Note that the shape factor decreases with increasing thickness. Therefore, the gasket should be as thin as possible to reduce relaxation. It must be thick enough, however, to permit adequate conformity. The clamp area should be as large as possible, consistent with seating stress requirements. Often designers reduce gasket width, thereby increasing gasket clamping stress to obtain better sealing. Remember, however, that this reduction might decrease the gasket's shape factor, resulting in higher relaxation in time.

2-6 ENVIRONMENTAL CONDITIONS

Many environmental conditions and factors influence the sealing performance of gaskets. Flange design details, in particular, are most important. Design details such as number, size, length, and spacing of clamping bolts; flange thickness and modulus; and surface finish, waviness, and flatness are important factors. Application specifics such as the medium being sealed, as well as the temperatures and pressures involved, also affect the gasket's sealing ability. The material must withstand corrosive attack of the confined medium. In particular, flange bowing is a most common type of problem associated with the sealing of a gasketed joint. The amount of bowing can be reduced by reducing the bolt spacing. For example, if the bolt spacing were cut in half, the bowing would be reduced to one-eighth of its original value [2-7]. Doubling of the flange thickness could also reduce bowing to one-eighth of its original value. A method of calculating the minimum stiffness required in a flange is available [2-8].

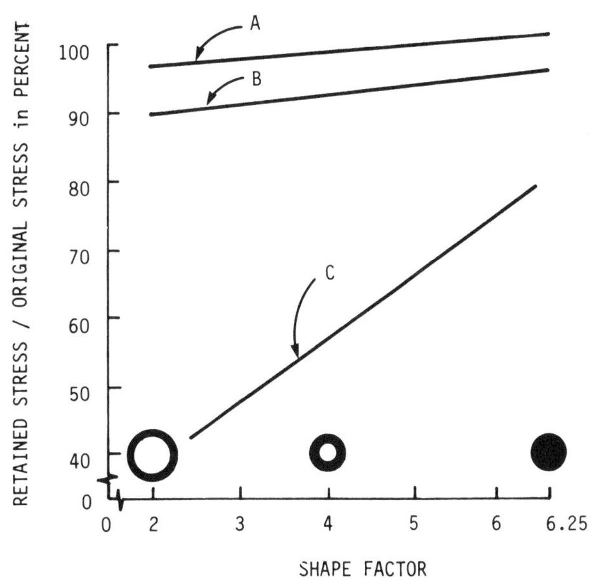

FIG 2-6 Retained stress for various gasket materials versus shape factor of the gasket. *A*, Asbestos fiber sheet; *B*, cellulose sheet; *C*, cork-rubber.

Different gasket materials and types require different surface finishes for optimum sealing. Soft gaskets such as rubber sheets can seal surface finishes in the vicinity of 500 microinches (μin) whereas some metallic gaskets may require finishes in the range of 32 μin for best sealing. Most gaskets, however, will seal adequately in the surface finish range of 63 to 125 μin with 90 to 110 μin being preferred. There are two main reasons for the surface finish differences: (1) The gasket must be able to conform to the roughness for surface sealing. (2) It must have adequate bite into the mating flange to create frictional forces to resist radial motion due to the internal pressure, thereby preventing blowout. In addition, elimination of the radial micromotion will result in maintaining the initial clampup sealing condition. Micromotion can result in localized fretting, and a leakage path may be created [2-9].

As a result of the complexity due to the wide variety of environmental conditions, some trial-and-error gasket designs for specific applications ensue. Understanding Sec. 2-7 will enable a designer to minimize the chance for leaks. Since it is so complex, however, adherence to the procedure will not ensure adequate performance in all cases. When inadequate gasket performance occurs, gasket manufacturers should be contacted for assistance.

2-7 GASKET DESIGN AND SELECTION PROCEDURE

2-7-1 Introduction

The first step in the selection of a gasket for sealing in a specific application is to choose a material that is both chemically compatible with the medium being sealed and thermally stable at the operating temperature of the application. The remainder of the selection procedure is associated with the minimum seating stress of the gasket and the internal pressure involved. In these regards two methods are proposed: the American Society of Mechanical Engineers (ASME) Code method and the simplified method proposed by Whalen.

2-7-2 ASME Code Procedure

The ASME Code for Pressure Vessels, Sec. VIII, Div. 1, App. 2, is the most commonly used design guide for gasketed joints. An integral part of this code centers on two factors: an m factor, called the *gasket material factor,* which is associated with the hydrostatic end force, and a y factor, which is the minimum seating stress associated with particular gasket material. The m factor is essentially a safety factor to increase the clamping load to such an amount that the hydrostatic end force does not unseat the gasket to the point of leakage. The factors were originally determined in 1937, and even though there have been objections to their specific values, these factors have remained essentially unchanged to date. The values are only suggestions and are not mandatory.

This method uses two basic equations for calculating required bolt load, and the larger of the two calculations is used for design. The first equation is associated with W_{m2} and is the required bolt load to initially seat the gasket:

$$W_{m2} = \pi bGy \tag{2-2}$$

The second equation states that the required bolt operating load must be sufficient to contain the hydrostatic end force and simultaneously maintain adequate compression on the gasket to ensure sealing:

Notes ▪ Drawings ▪ Ideas

$$W_{m1} = \frac{\pi}{4} G^2P + 2b\pi GmP \tag{2-3}$$

where W_{m1} = required bolt load for maximum operating or working conditions, lb
W_{m2} = required initial bolt load at atmospheric temperature conditions without internal pressure, lb
G = diameter at location of gasket load reaction generally defined as follows: When $b_0 \leq \frac{1}{4}$ in, G = mean diameter of gasket contact face, in; when $b_0 > \frac{1}{4}$ in, G = outside diameter of gasket contact face less $2b$, in
P = maximum allowable working pressure, psi
b = effective gasket or joint-contact-surface seating width, in
$2b$ = effective gasket or joint-contact-surface pressure width, in
b_0 = basic gasket seating width per Table 2-4 (the table defines b_0 in terms of flange finish and type of gasket, usually from one-half to one-fourth gasket contact width)
m = gasket factor per Table 2-3 (the table shows m for different types and thicknesses of gaskets ranging from 0.5 to 6.5)
y = gasket or joint-contact-surface unit seating load, psi (per Table 2-3, which shows values from 0 to 26 000 psi)

Tables 2-3 and 2-4 are reprints of Tables 2-5-1 and 2-5-2 of the 1980 ASME Code [2-10].

To determine bolt diameter based on required load and a specified torque for the grade of bolt, the following is used:

$$W_b = 0.17DT \qquad \text{(for lubricated bolts)} \tag{2-4}$$

or

$$W_b = 0.2DT \qquad \text{(for unlubricated bolts)} \tag{2-5}$$

where W_b = load per bolt, lb
D = bolt diameter, in
T = torque for grade of bolt selected, lb·in

Note that W_b is the load per bolt and must be multiplied by the number of bolts to obtain total bolt load.

To determine the bolt diameter based on the required load and the allowable bolt stress for a given grade of bolt, use

$$W_b = \sigma_b A_b \tag{2-6}$$

where W_b = load per bolt, lb
σ_b = allowable bolt stress for grade of bolt selected, psi
A_b = minimum cross-sectional area of bolt, in^2

2-7-3 Simplified Procedure

A simpler method of calculation has been suggested by Whalen [2-11]. This method is also based on the seating stress σ_g on the gasket, as shown in Table 2-5 (page 65), and on the hydrostatic end force involved in the application. Basically, Whalen's equations accomplish the same thing as the Code, but they are simplified since they use the full gasket contact width, regardless of the flange width and the surface finish of the sealing faces.

Notes ▪ Drawings ▪ Ideas

TABLE 2-3 Gasket Materials and Contact Facings†
Gasket Factors *m* for Operating Conditions and Minimum Design Seating Stress *y*

Gasket material	Gasket factor m	Minimum design seating stress *y*, psi	Sketches	Facing sketch and column to be used from Table 2-4
Self-energizing types (O-rings, metallic, elastomer, other gasket types considered as self-sealing)	0	0		
Elastomers without fabric or high percentage of asbestos fiber:			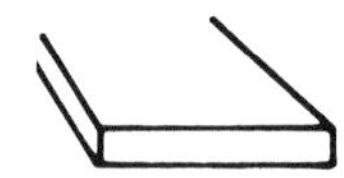	(1a), (1b), (1c), (1d), (4), (5); column II
Below 75A Shore Durometer	0.50	0		
75A or higher Shore Durometer	1.00	200		
Asbestos with suitable binder for operating conditions:			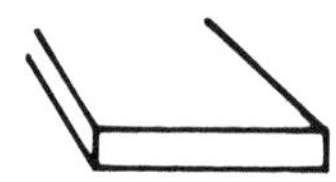	(1a), (1b), (1c), (1d), (4), (5); column II
$\frac{1}{8}$ in thick	2.00	1 600		
$\frac{1}{16}$ in thick	2.75	3 700		
$\frac{1}{32}$ in thick	3.50	6 500		
Elastomers with cotton fabric insertion	1.25	400	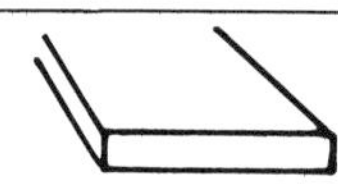	(1a), (1b), (1c), (1d), (4), (5); column II
Elastomers with asbestos fabric insertion (with or without wire reinforcement):			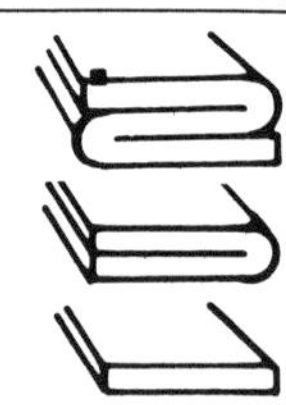	(1a), (1b), (1c), (1d), (4), (5); column II
3-ply	2.25	2 200		
2-ply	2.50	2 900		
1-ply	2.75	3 700		
Vegetable fiber	1.75	1 100	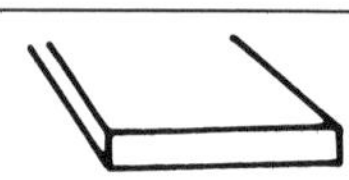	(1a), (1b), (1c), (1d), (4), (5); column II
Spiral wound metal, asbestos-filled:			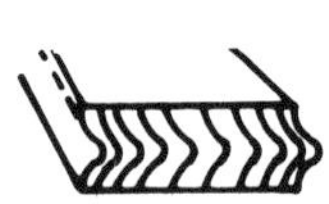	(1a), (1b); column II
Carbon	2.50	10 000		
Stainless or Monel	3.00	10 000		
Corrugated metal, asbestos inserted or corrugated metal, jacketed asbestos-filled:			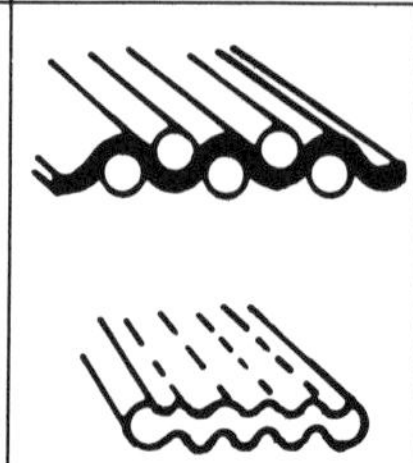	(1a), (1b); column II
Soft aluminum	2.50	2 900		
Soft copper or brass	2.75	3 700		
Iron or soft steel	3.00	4 500		
Monel or 4–6% chrome	3.25	5 500		
Stainless steels	3.50	6 500		

TABLE 2-3 Gasket Materials and Contact Facings†
Gasket Factors *m* for Operating Conditions and Minimum Design Seating Stress *y* (*Continued*)

Gasket material	Gasket factor m	Minimum design seating stress y, psi	Sketches	Facing sketch and column to be used from Table 2-4
Corrugated Metal:				
Soft aluminum	2.75	3 700		(1a), (1b), (1c), (1d); column II
Soft copper or brass	3.00	4 500		
Iron or soft steel	3.25	5 500		
Monel or 4–6% chrome	3.50	6 500		
Stainless steels	3.75	7 600		
Flat metal, jacketed asbestos-filled:			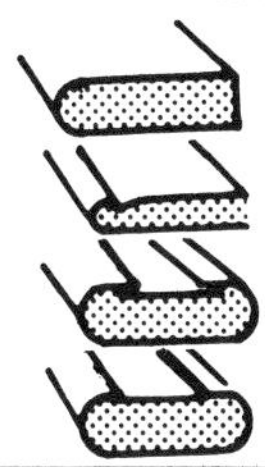	
Soft aluminum	3.25	5 500		(1a), (1b), (1c),‡ (1d),‡ (2)‡; column II
Soft copper or brass	3.50	6 500		
Iron or soft steel	3.75	7 600		
Monel or 4–6% chrome	3.50	8 000		
	3.75	9 000		
Stainless steels	3.75	9 000		
Grooved metal:				
Soft aluminum	3.25	5 500		(1a), (1b), (1c), (1d), (2), (3); column II
Soft copper or brass	3.50	6 500		
Iron or soft steel	3.75	7 600		
Monel or 4–6% chrome	3.75	9 000		
Stainless steels	4.25	10 100		
Solid flat metal:				
Soft aluminum	4.00	8 800		(1a), (1b), (1c), (1d), (2), (3), (4), (5); column I
Soft copper or brass	4.75	13 000		
Iron or soft steel	5.50	18 000		
Monel or 4–6% chrome	6.00	21 800		
Stainless steels	6.50	26 000		
Ring joint:				
Iron or soft steel	5.50	18 000	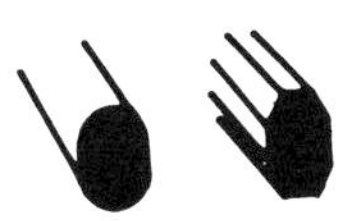	(6); column I
Monel or 4–6% chrome	6.00	21 800		
Stainless steels	6.50	26 000		

†This table gives a list of many commonly used gasket materials and contact facings with suggested design values of m and y that have generally proved satisfactory in actual service when using effective gasket seating width b given in Table 2-4. The design values and other details given in this table are only suggested and are not mandatory.

‡The surface of a gasket having a lap should not be against the nubbin.

TABLE 2-4 Effective Gasket Width†

	Facing sketch (exaggerated)		Basic gasket seating width b_o: Column I	Basic gasket seating width b_o: Column II
(1a)	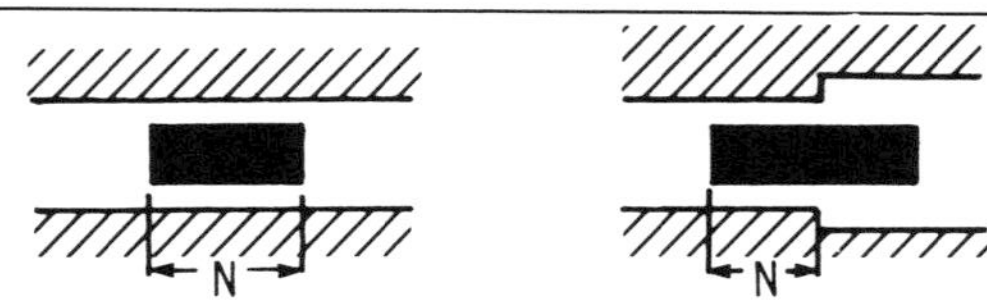		$\frac{N}{2}$	$\frac{N}{2}$
(1b)‡	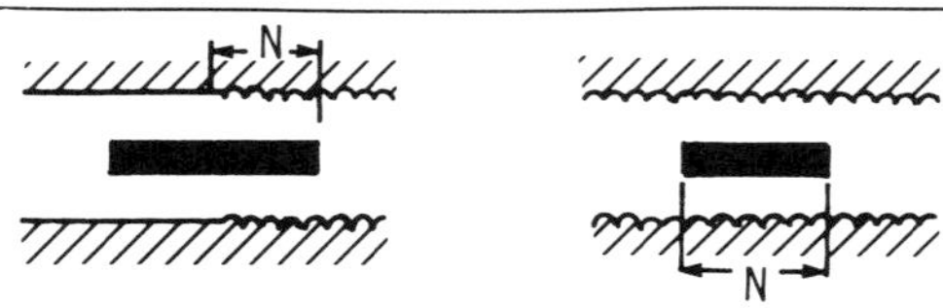			
(1c)	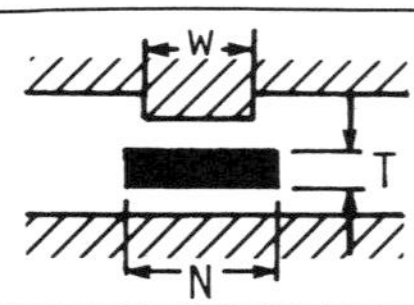	$w \le N$	$\frac{w+T}{2}\left(\frac{w+N}{4}\text{ max.}\right)$	$\frac{w+T}{2}\left(\frac{w+N}{4}\text{ max.}\right)$
(1d)‡	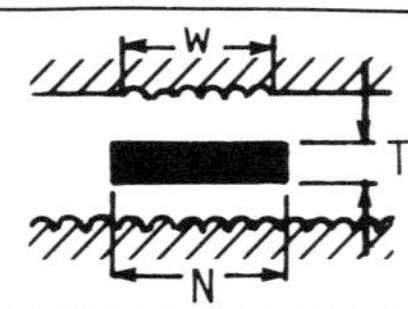	$w \le N$		
(2) 1/64″ Nubbin	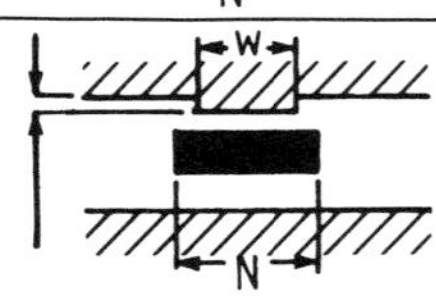	$w \le \frac{N}{2}$	$\frac{w+N}{4}$	$\frac{w+3N}{8}$

(3) ¹⁄₆₄″ Nubbin	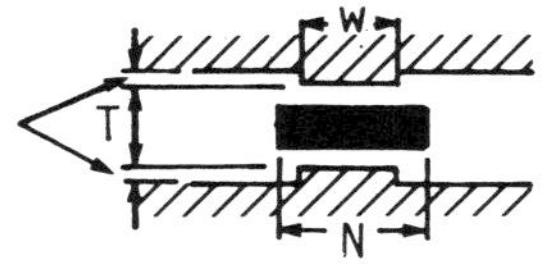	$w \le \frac{N}{2}$	$\frac{N}{4}$	$\frac{3N}{8}$
(4)‡	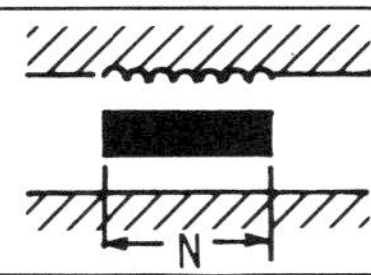		$\frac{3N}{8}$	$\frac{7N}{16}$
(5)‡	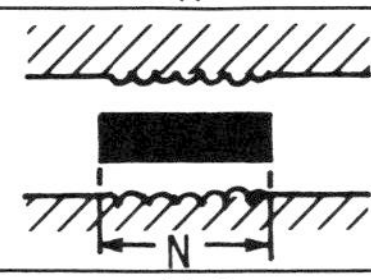		$\frac{N}{4}$	$\frac{3N}{4}$
(6)	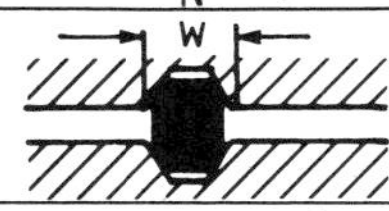		$\frac{w}{8}$	

S81 Effective gasket seating width b:

$b = b_o$ when $b_o \le \frac{1}{4}$ in $\qquad b = 0.5\sqrt{b_o}$ when $b_o > \frac{1}{4}$ in

Location of gasket load reaction:

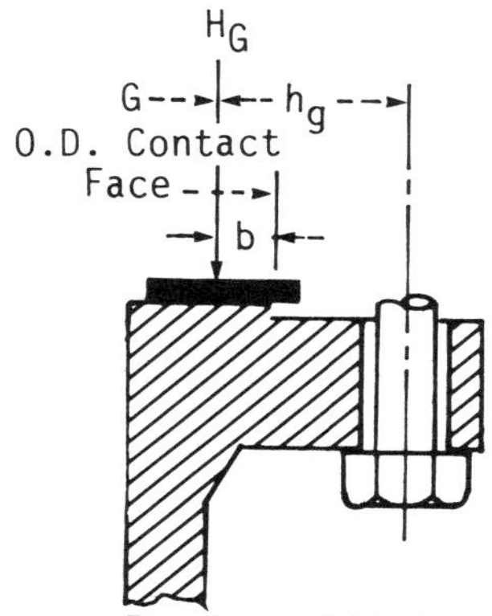

For b_o > 1/4 in.

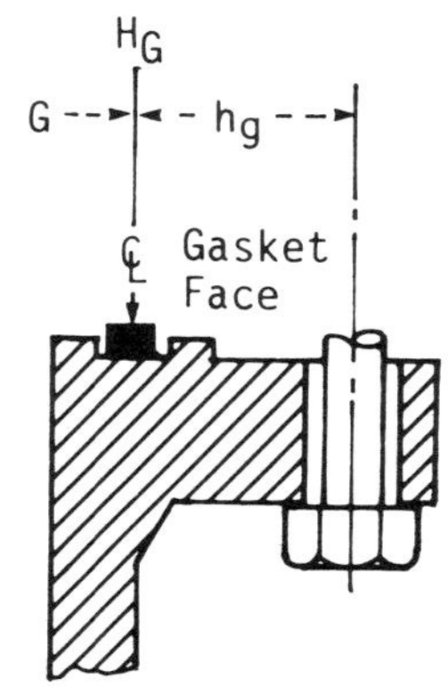

For $b_o \leqq$ 1/4 in.

†The gasket factors listed apply only to flanged joints in which the gasket is contained entirely within the inner edges of the bolt holes.
‡Where separations do not exceed $\frac{1}{64}$-in-depth and $\frac{1}{32}$-in-width spacing, sketches (1b) and (1d) shall be used.

Notes ▪ Drawings ▪ Ideas

TABLE 2-5 Minimum Recommended Seating Stresses for Various Gasket Materials

	Material	Gasket type	Minimum seating stress range (S_g), psi†
Nonmetallic	Asbestos fiber sheet	Flat	
	$\frac{1}{8}$ in thick		1400 to 1600
	$\frac{1}{16}$ in thich		3500 to 3700
	$\frac{1}{32}$ in thick		6000 to 6500
	Asbestos fiber sheet $\frac{1}{32}$ in thick	Flat with rubber beads	1000 to 1500 lb/in on beads
	Asbestos fiber sheet $\frac{1}{32}$ in thick	Flat with metal grommet	3000 to 4000 lb/in on grommet
	Asbestos fiber sheet $\frac{1}{32}$ in thick	Flat with metal grommet and metal wire	2000 to 3000 lb/in on wire
	Cellulose fiber sheet	Flat	750 to 1100
	Cork composition	Flat	400 to 500
	Cork-rubber	Flat	200 to 300
	Fluorocarbon (TFE)	Flat	
	$\frac{1}{8}$ in thick		1500 to 1700
	$\frac{1}{16}$ in thick		3500 to 3800
	$\frac{1}{32}$ in thick		6200 to 6500
	Nonasbestos fiber sheets (glass, carbon, aramid, and ceramics)	Flat	1500 to 3000 depending on composition
	Rubber	Flat	100 to 200
	Rubber with fabric or metal reinforcement	Flat with reinforcement	300 to 500
Metallic	Aluminum	Flat	10 000 to 20 000
	Copper	Flat	15 000 to 45 000 depending on hardness
	Carbon steel	Flat	30 000 to 70 000 depending on alloy and hardness
	Stainless steel	Flat	35 000 to 95 000 depending on alloy and hardness
	Aluminum (soft)	Corrugated	1000 to 3700
	Copper (soft)	Corrugated	2500 to 4500
	Carbon steel (soft)	Corrugated	3500 to 5500
	Stainless steel	Corrugated	6000 to 8000
	Aluminum	Profile	25 000
	Copper	Profile	35 000
	Carbon steel	Profile	55 000
	Stainless steel	Profile	75 000
Jacketed metal-asbestos	Aluminum	Plain	2 500
	Copper	Plain	4 000
	Carbon steel	Plain	6 000
	Stainless steel	Plain	10 000
	Aluminum	Corrugated	2000
	Copper	Corrugated	2500
	Carbon steel	Corrugated	3000
	Stainless steel	Corrugated	4000
	Stainless steel	Spiral-wound	3000 to 30 000

†Stresses in pounds per square inch except where otherwise noted.

Notes ▪ Drawings ▪ Ideas

TABLE 2-6 Typical Gasket Designs and Descriptions

Type	Cross section	Comments
Flat		Basic form. Available in wide variety of materials. Easily fabricated into different shapes.
Reinforced		Fabric- or metal-reinforced. Improves torque retention and blowout resistance of flat types. Reinforced type can be corrugated.
Flat with rubber beads		Rubber beads located on flat or reinforced material afford high unit sealing pressure and high degree of conformability.
Flat with metal grommet		Metal grommet affords protection to base material from medium and provides high unit sealing stress. Soft metal wires can be put under grommet for higher unit sealing stress.
Plain metal jacket		Basic sandwich type. Filler is compressible. Metal affords protection to filler on one edge and across surfaces.
Corrugated or embossed	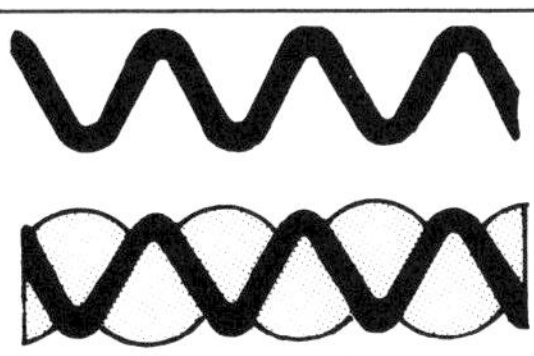	Corrugations provide for increased sealing pressure and higher conformability. Primarily circular. Corrugations can be filled with soft filler.
Profile	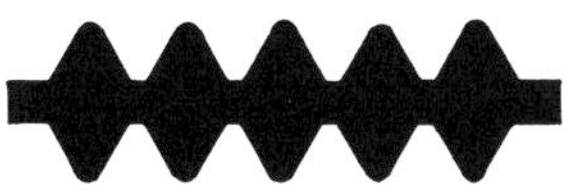	Multiple sealing surfaces. Seating stress decreases with increase in pitch. Wide varieties of designs are available.
Spiral-wound		Interleaving pattern of metal and filler. Ratio of metal to filler can be varied to meet demands of different applications.

This method is based on the total bolt load F_b being sufficient to

1. Seat the gasket material into the flange surface
2. Prevent the hydrostatic end force from unseating the gasket to the point of leakage

In the first case, Table 2-5 lists a range of seating-stress values. The ranges shown were found in a search of the literature on gasket seating stresses. Gasket suppliers can be contacted to confirm these values.

Table 2-6 depicts various gasket types and comments on them. As a starting point in the design procedure, the mean value of σ_g could be used. Then, depending on the severity of the application and/or the safety factor desired, the upper and lower figures could be utilized.

Two equations are associated with this procedure. The first is

$$F_b = \sigma_g A_g \tag{2-7}$$

where F_b = total bolt load, lb
σ_g = gasket seating stress, psi (from Table 2-5)
A_g = gasket contact area, in^2

This equation states that the total bolt load must be sufficient to seat the gasket when the hydrostatic end force is not a major factor. The second equation associated with the hydrostatic end force is

$$F_b = KP_t A_m \tag{2-8}$$

where P_t = test pressure or internal pressure if no test pressure is used
A_m = hydrostatic area on which internal pressure acts (normally based on gasket's middiameter)
K = safety factor (from Table 2-7)

The safety factors K from Table 2-7 are based on the joint conditions and operating conditions but not on the gasket type or flange surface finish. They are similar to the m factors in the ASME Code. Equation (2-8) states that the total bolt load must be more than enough to overcome the hydrostatic end force. The middiameter is used in A_m since testing has shown that just prior to leakage, the internal pressure acts up to the middiameter of the gasket.

TABLE 2-7 Safety Factors for Gasketed Joints

K factor	When to apply
1.2 to 1.4	For minimum-weight applications where all installation factors (bolt lubrication, tension, parallel seating, etc.) are carefully controlled; ambient to 250°F (121°C) temperature applications; where adequate proof pressure is applied.
1.5 to 2.5	For most normal designs where weight is not a major factor, vibration is moderate and temperatures do not exceed 750°F (399°C). Use high end of range where bolts are not lubricated.
2.6 to 4.0	For cases of extreme fluctuations in pressure, temperature, or vibration; where no test pressure is applied; or where uniform bolt tension is difficult to ensure.

Notes ▪ Drawings ▪ Ideas

After the desired gasket has been selected, the minimum seating stress, as given in Table 2-5, is used to calculate the total bolt load required by Eq. (2-7). Then the bolt load required to ensure that the hydrostatic end force does not unseat the gasket is calculated from Eq. (2-8). The total bolt load F_b calculated by Eq. (2-7) must be greater than the bolt load calculated in Eq. (2-8). If it is not, then the gasket design must be changed, the gasket's area must be reduced, or the total bolt load must be increased.

Both the ASME procedure and the simplified procedure are associated with gasketed joints which have rigid, usually cast-iron flanges, have high clamp loads, and generally contain high pressures. A great many gasketed joints have stamped-metal covers and splash or very low fluid pressure. In these cases, the procedures do not apply, and the compression and stress distribution discussed next should be considered by the designer.

2-8 GASKET COMPRESSION AND STRESS-DISTRIBUTION TESTING

After a gasket has been selected and designed for a particular application, two simple tests can be performed to determine the gasket's compressed thickness and stress distribution. Inadequate compression or nonuniform stress distribution could result in a leaking joint. The tests can be performed to check for these possibilities and permit correction to ensure leaktight joints.

1. *Lead Pellet Test* In this test, lead pellets are used to accurately indicate the compressed thicknesses of a gasketed joint. The pellets, commonly called *lead shot,* are available from local gun supply stores. A size approximately twice the thickness of the gasket should be used. Lead solid-core solder can also be used if desired with the same size requirements. Pellets or solder are particularly well suited for doing this test as they exhibit no recovery after compression whereas the actual gasketing material will almost always exhibit some recovery. The degree of nonuniform loading, flange bowing, or distortion will be indicated by the variations in the gasket's compressed thickness.

To begin, the original thickness of the gasket is measured and recorded at uniformly selected points across the gasket. At or near these points, holes are punched or drilled through the gasket. Care should be taken to remove any burrs. The punched holes should be approximately $1\frac{1}{2}$ times the pellet diameter.

Then the gasket is mounted on the flange. A small amount of grease can be put in the punched holes to hold the lead pellets, if required. The pellets are mounted in the grease, and the mating flange is located and torqued to specifications.

Upon careful disassembly of the flange and removal of the pellets, their thicknesses are measured, recorded, and analyzed. Comparison of the pellets' compressed thicknesses to the gasket's stress-compression characteristics permits the desired stress-distribution analysis.

2. *NCR Paper Test* This test utilizes no-carbon-required (NCR) paper for visual determination of the stress distribution on a gasket. NCR paper is a pressure-sensitive, color-reactive paper. The intensity of color is proportional to the stress imposed on the paper, which is the same stress as is on the gasket.

NCR paper is available from the National Cash Register Company, the 3M Company, and other paper companies. Various grades are available, but the medium grade is usually chosen. Some papers are only one sheet while others are composed of two sheets. Either type can be used.

To begin, the bolt holes are pierced in a piece of the impression paper. The pierced holes in the paper are made slightly larger than the bolts. The paper is placed on the flange, and the mating flange is assembled per torque specifications.

When you are using the two-piece carbonless paper, make sure to keep the two papers oriented to each other as they were purchased; otherwise, no impression may result. Upon torquing, the impression is made on the paper. The flange is removed, and the impression is inspected for stress distribution. A judgment of the gasket's sealing ability can now be made. Further analysis can be done by calibrating the load versus color intensity of the paper. Various known stresses can be applied to the paper and the resulting color impressions identified. The impressions can be compared to the test sample, and then the stress on the sample can be determined.

In both the lead pellet and NCR paper tests, gasket manufacturers can be contacted for further interpretation of the results and more detailed analysis.

2-9 INSTALLATION SPECIFICATIONS

An installation is only as good as its gasket; likewise, a gasket is only as good as its installation [2-1]. The following are some recommendations associated with gasket installation:

1. Be sure that mating surfaces are clean and in specification with regard to finish, flatness, and waviness.
2. Check gasket for damage before installing it.
3. Make certain the gasket fits the application.
4. Specify lubricated bolts. Bolt threads and the underside of the bolt head should be lubricated.
5. Specify the torque level.
6. Specify the torquing sequence. In addition to the sequence, two or three stages of torque before reaching specified level are recommended.

PART 2 SEALS

R. BRUCE HOPKINS

2-10 ELASTOMERIC SEAL RINGS

Seal rings of the O-ring type are used as both static and dynamic seals. Static seals serve the same purpose as gaskets; that is, they provide a seal between two members that are not intended to undergo relative motion. Dynamic seals, however, are used where rotating or reciprocating motion is intended to occur.

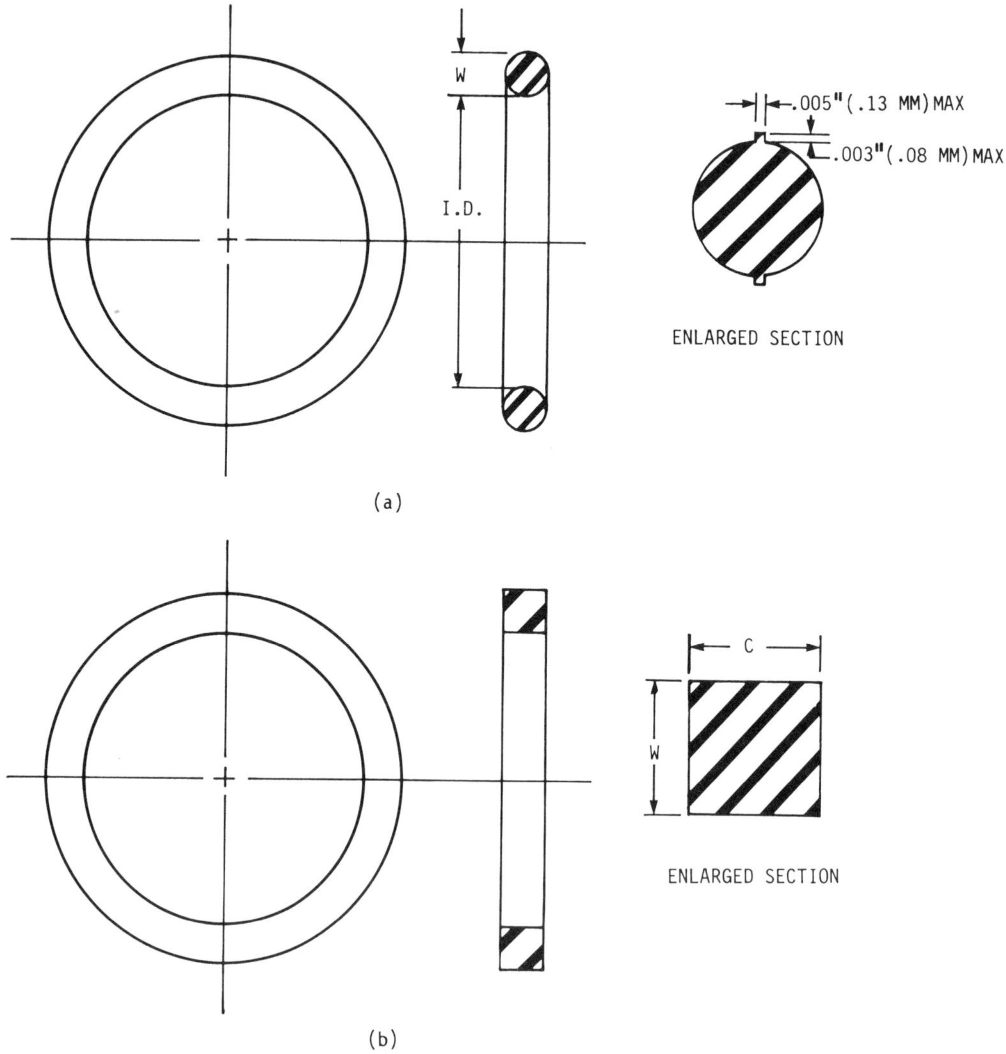

FIG. 2-7 Seal rings. (*a*) O-ring; (*b*) rectangular-section ring.

O-rings are molded to the size of the elastomeric material with a circular cross section, as shown in Fig. 2-7*a*. The *size* is designated by the cross-sectional diameter *w* and the nominal inside diameter (ID). The standard sizes specified in SAE J120a are summarized in Table 2-8. The first size number in a group is associated with the minimum inside diameter, and the last size number is associated with the maximum inside diameter. Some manufacturers provide additional sizes that extend the range of inside diameters for a particular cross-section size. The nominal inside diameters were selected to provide dynamic seals in cylinder bores dimensioned in inches and common fractions of inches. Either SAE J120a or manufacturer's recommendations should be consulted to obtain the recommended compression of the cross section. The compression is different for static and dynamic applications.

The rectangular-section ring in Fig. 2-7*b* is manufactured by cutting lengths from a tube of molded material. The standard sizes listed in SAE J120a are summarized in Table 2-9. The first size number in a group is associated with the minimum inside diameter, and the last size number is associated with the maximum inside

TABLE 2-8 Standard Sizes of O-Rings

Size no.	w, in	Actual ID, in	
		Minimum	Maximum
006 to 045	0.070 ± 0.003	0.114 ± 0.005	3.989 ± 0.015
110 to 163	0.103 ± 0.003	0.362 ± 0.005	5.987 ± 0.023
210 to 281	0.139 ± 0.004	0.734 ± 0.006	14.984 ± 0.060
325 to 349	0.210 ± 0.005	1.475 ± 0.010	4.475 ± 0.010
425 to 460	0.275 ± 0.006	4.475 ± 0.015	15.475 ± 0.015

diameter. Rectangular-section rings are suitable for static applications with pressures up to 1500 psi [10.3 newtons per square millimeter (N/mm^2)].

The standard shape of groove for sealing rings is shown in Fig. 2-8. The actual groove dimensions depend on the type and size of the seal ring cross section and the nature of the application. Recommended groove dimensions are provided in SAE J120a and in the manufacturer's literature. Because elastomeric materials are almost incompressible, it is necessary to provide sufficient volume for the seal ring in the groove. The recommended groove dimensions do so.

For static seals, a finish on surfaces contacted by the seal ring that is rougher than 32 μin (0.8 μm) may lead to leakage. Because rough finishes accelerate seal wear in dynamic seals, a surface finish of 5 to 16μin (0.13 to 0.4 μm) is preferred. Friction is reduced with the smoother finish, but surfaces smoother than 5 μin (0.13 μm) may not be satisfactory for reciprocating motion.

A static seal ring application in which the joint is subject to internal pressure only is shown in Fig. 2-9*a*. The groove design in Fig. 2-9*b* is for a joint subject to external pressure or internal vacuum only. It is generally advisable in these applications to use as large a seal ring cross section as possible because the tolerance on the groove depth is greater with larger cross sections. This requires less precise machining and tends to reduce manufacturing costs.

O-rings are also used as static seals for hydraulic tube fittings that are screwed into tapped holes. Recommended maching dimensions are provided in SAE 514 (April 1980).

TABLE 2-9 Standard Sizes of Rectangular-Section Rings

Size no.	w, in	c, in	Actual ID, in	
			Minimum	Maximum
R006 to R045	0.066 ± 0.004	0.066 ± 0.003	0.114 ± 0.005	3.989 ± 0.015
R110 to R163	0.099 ± 0.004	0.099 ± 0.003	0.362 ± 0.005	5.987 ± 0.023
R210 to R281	0.134 ± 0.004	0.134 ± 0.004	0.734 ± 0.006	14.984 ± 0.060
R325 to R349	0.203 ± 0.005	0.203 ± 0.005	1.475 ± 0.010	4.475 ± 0.015
R425 to R460	0.265 ± 0.005	0.265 ± 0.005	4.475 ± 0.015	15.475 ± 0.030

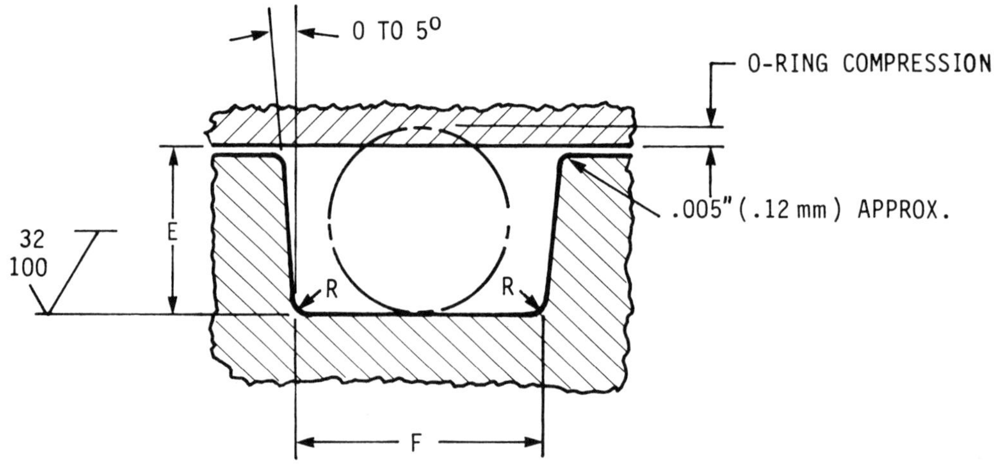

FIG. 2-8 Shape of groove for seal rings.

Elastomeric sealing rings are most commonly made of nitrile (Buna N) compounds. These compounds are low in cost and are compatible with alcohol, gasoline, hydraulic fluids, lubricating oils, and water. They also are suitable for temperatures ranging from −67 to 257°F (−55 to 125°C). For resistance to higher temperatures or compatibility with other fluids, other compounds are employed. Among these compounds are butyl, ethylene propylene, neoprene, fluorocarbon, silicone, and polyurethane.

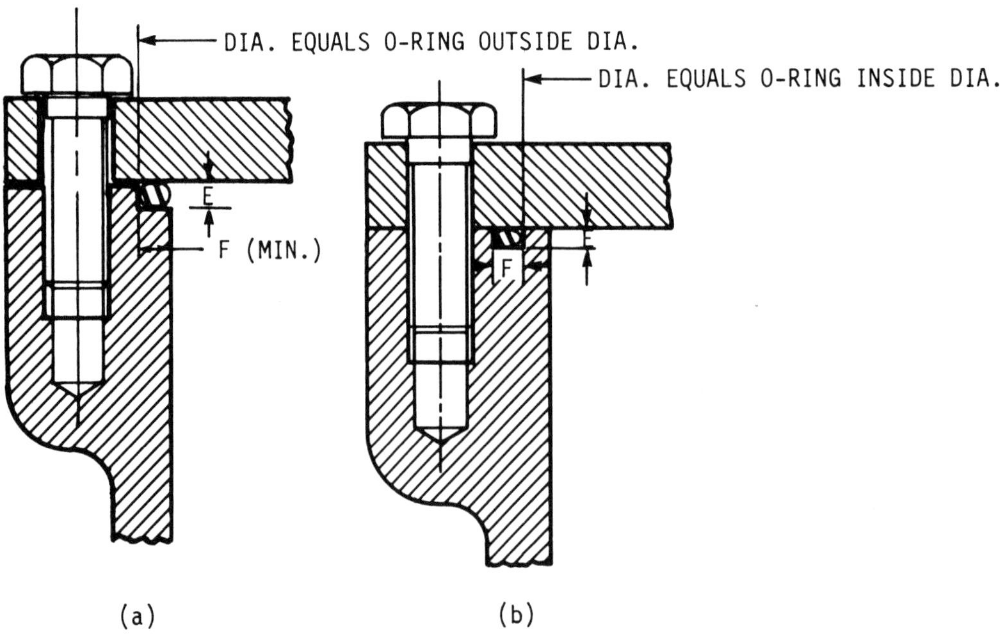

FIG. 2-9 Static O-ring seals. (*a*) Joint subject to internal pressure only; (*b*) joint subject to external pressure.

Notes ▪ Drawings ▪ Ideas

2-11 SEALS FOR ROTARY MOTION

Seals are required on rotating shafts to retain working fluids, retain lubricants, and to exclude dirt. The selection of a seal type depends on fluid pressure, shaft speed, and whether any leakage can be permitted. There are many variations of the basic seal types that are available from various manufacturers.

2-11-1 O-Rings

Attempts to use O-rings as seals for rotating shafts have not always been successful because the elastomers shrink when heated. If an O-ring is under tension, friction between the ring and shaft generates heat that makes the ring shrink. Contraction of the ring creates additional heat, and failure occurs rapidly.

O-rings have been used successfully on rotating shafts when they are installed under compression by using a smaller-than-normal groove diameter in the housing. Satisfactory life can then be obtained at shaft speeds up to 750 feet per minute (ft/min) [3.8 meters per second (m/s)] and sealed pressures up to 200 psi (1.38 N/mm^2). Recommended O-ring cross sections are 0.139 in (3.53 mm) for speeds up to 400 ft/min (2.0 m/s), 0.103 in (2.62 mm) for speeds from 400 to 600 ft/min (2.0 to 3.0 m/s), and 0.070 in (1.78 mm) for speeds exceeding 600 ft/min (3.0 m/s) [2-13].

2-11-2 Radial Lip Seals

A section through a radial lip seal is shown in Fig. 2-10. This type is used primarily for retention of lubricants and exclusion of dirt. It is suited for conditions of low lubricant pressure, moderate shaft speeds, less-than-severe environmental conditions, and where slight leakage may be permitted. Radial lip seals are compact, effective, inexpensive, and easily installed.

The outer case is held in the bearing housing by an interference fit. The garter spring provides a uniform radial force to maintain contact between the elastomeric

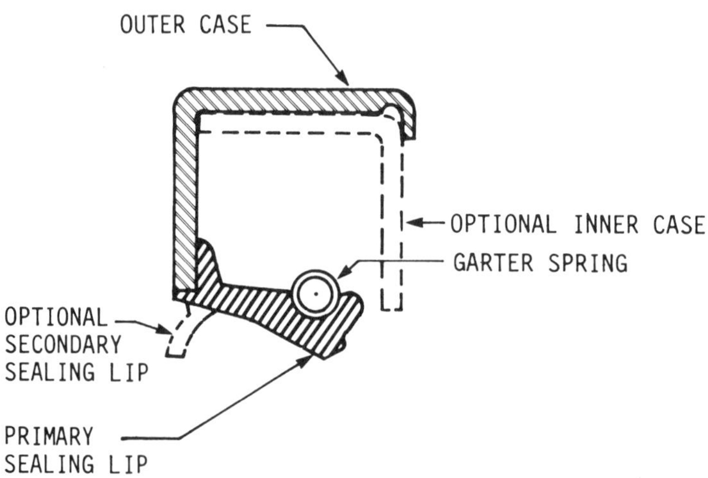

FIG. 2-10 Cross section of radial lip seal.

sealing lips and the shaft. Lubricant leakage is reduced when hydrodynamic sealing lips are used. Such lips have very shallow grooves molded into the primary sealing lip to pump lubricant out of the contact area. Hydrodynamic sealing lips are manufactured for rotation in one direction only or for rotation in either direction.

Sealing lips are most commonly made of nitrile (Buna N) rubber compounds because of their compatibility with greases, lubricating oils, and hydraulic fluids. The nitrile compounds have poor to fair compatibility with extreme-pressure (EP) additives used in some gear lubricants. A polyacrylate or fluoroelastomer compound is a better choice with EP lubricants.

Radial lip seal terminology is presented in SAE J111c, and recommendations for applications are made in SAE J946d. One of the purposes of the secondary sealing lip shown in Fig. 2-10 is to exclude dust. That lip, however, leads to higher seal temperatures because of the additional friction, and the higher temperatures lead to earlier seal failure. Dual lip seals are recommended for shaft speeds exceeding 150 ft/min (0.76 m/s) [2-14].

A minimum hardness of Rockwell C 30 is recommended for the portions of shafts that contact the sealing lips in order to prevent scoring of the shaft. If the shaft may be damaged in handling, a minimum hardness of Rockwell C 45 will provide protection against damage. A hard surface can be provided for soft shafts by use of a hardened wear sleeve of thin steel that is held in place by an interference fit.

Radial lip seals function best with carbon-, alloy-, or stainless-steel shafts or nickel-plated surfaces. Use with aluminum alloys, brass, bronze, magnesium, zinc, or similar metals is not recommended. Shaft surface texture should be in the range of 10 to 20 μin (0.25 to 0.50 μm). This condition can best be met by plunge-grinding.

This type of seal is limited to sealing pressures of 3 psig (0.02 N/mm^2 gauge) at shaft speeds exceeding 2000 ft/min (10.2 m/s) and 7 psig (0.05 N/mm^2 gauge) at speeds up to 1000 ft/min (5.1 m/s). When pressures exceed these limits, a mechanical face seal is preferable.

2-11-3 Face Seals

Whereas a radial lip seal contacts the shaft circumference, a face seal acts against a surface perpendicular to the shaft axis. The seal may be mounted in a housing and seal against a shoulder or collar on the shaft. The seal also may be mounted on the shaft and seal against a surface on the housing. Some elastomeric face sealing elements are loaded by mechanical springs while others are not.

Elastomeric face seals (Fig. 2-11*a*) are used to retain lubricants. For high-speed applications the lubricant should be on the side where centrifugal force throws the lubricant into the sealing area. These seals have the disadvantage of requiring rather precise location with respect to the sealing surface in order to provide the proper force on the seal. The sealing surface must be flat and smooth with a surface finish of 10 to 20 μin (0.25 to 0.50 μm). Less rigid control of surface flatness is required for low speeds.

Mechanical face seals are used in situations where a radial lip seal or elastomeric face seal will not be satisfactory. Abrasive conditions, such as those encountered by earth-moving machinery and mining machinery, may dictate the use of a mechanical face seal. A mechanical seal is frequently used in the automotive coolant pump (Fig. 2-11*b*) in which the pressure is relatively low. Here, the stationary spring-loaded seal ring contacts a flat surface on the rotor.

If the seal in Fig. 2-11*b* were used with high-pressure fluid, a high axial sealing force would result. Friction between the rotor and stator could possibly cause over-

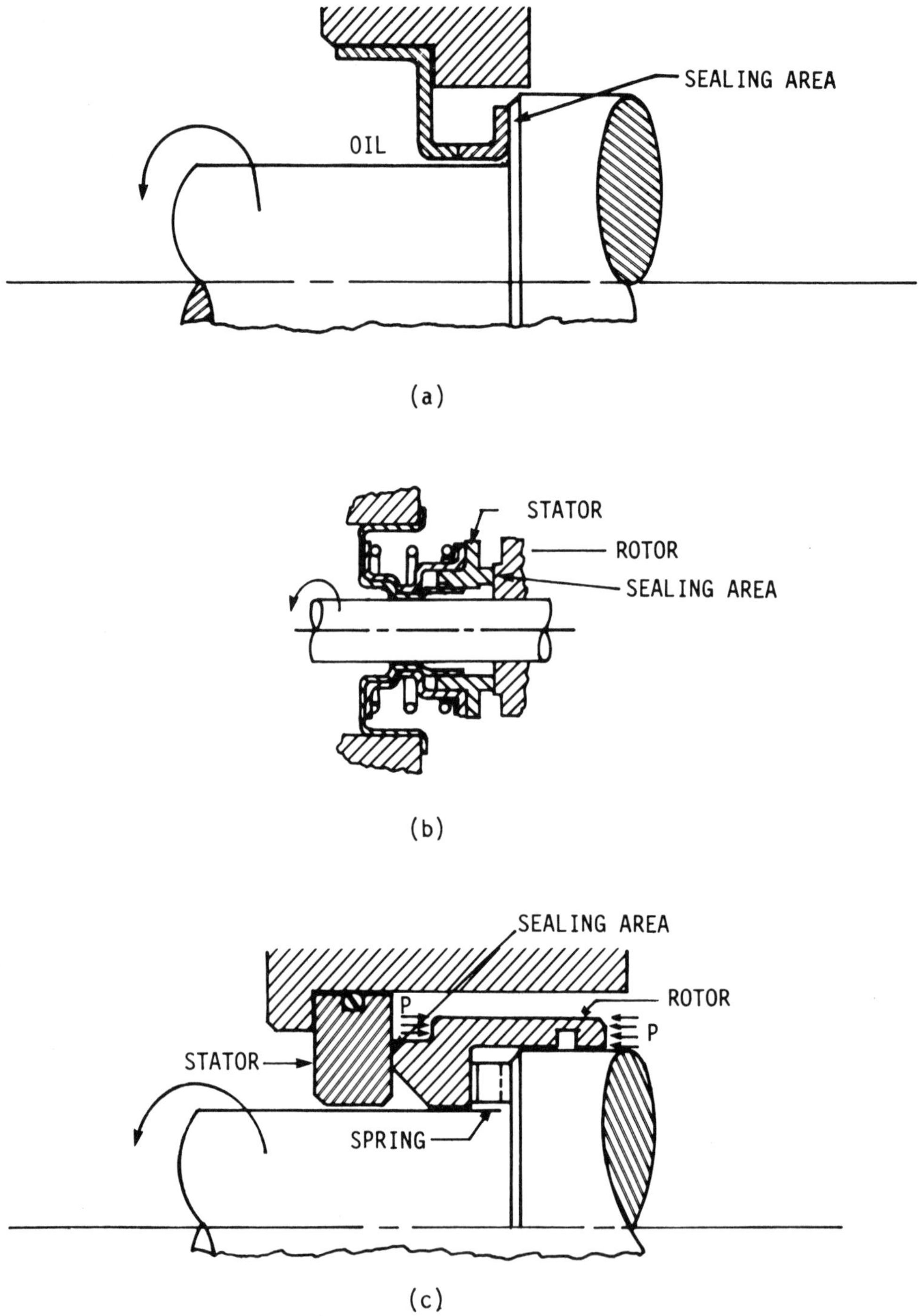

FIG. 2-11 Face seals. (*a*) Housing-mounted elastomeric seal; (*b*) mechanical seal for engine coolant pump; (*c*) balanced mechanical seal.

Notes ▪ Drawings ▪ Ideas

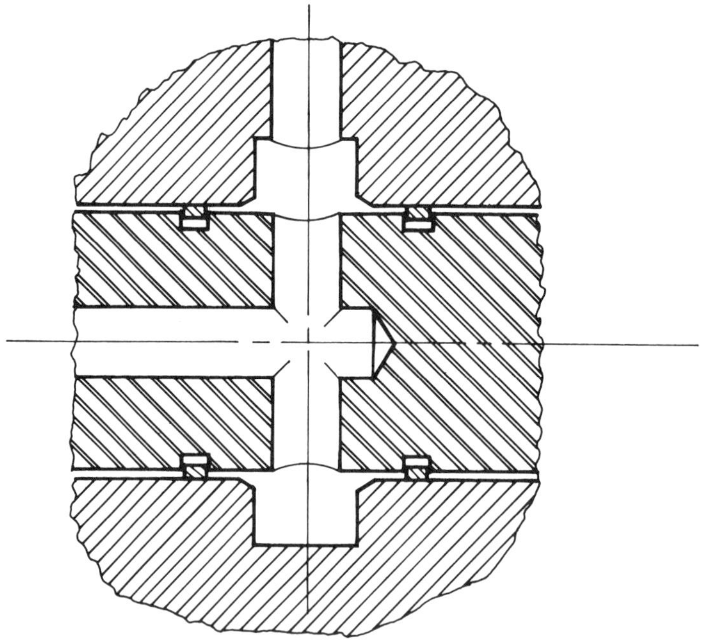

FIG. 2-12 Metal seal ring application on a rotating shaft.

heating of the seal elements. Consequently, a balanced mechanical face seal, such as in Fig. 2-11*c*, is used for higher pressures. These seals are proportioned so that much of the force due to pressure is balanced. This leaves a small net force to provide contact between rotor and stator.

2-11-4 Metal Sealing Rings

Cast-iron sealing rings are used in hydraulic applications where oil must be introduced into a rotating shaft (Fig. 2-12). A typical application is to operate a clutch in an automatic transmission for a motor vehicle. Ring cross-sectional dimensions are similar to those for engine piston rings of the same outside diameter. Information on designing to accommodate these rings is provided in SAE J281 (Sept. 1980) and SAE J1236 (Sept. 1980).

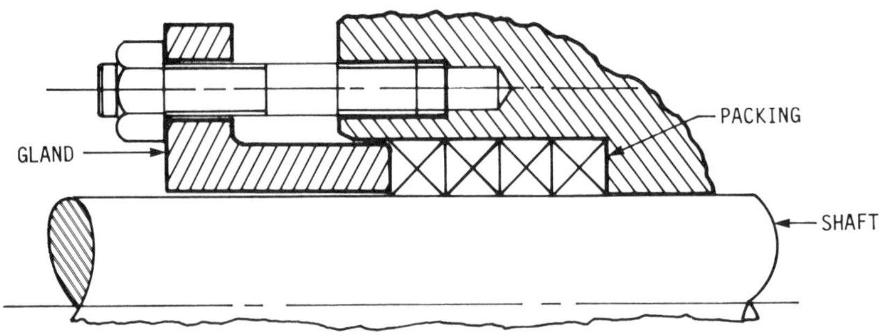

FIG. 2-13 Stuffing box for a rotating shaft.

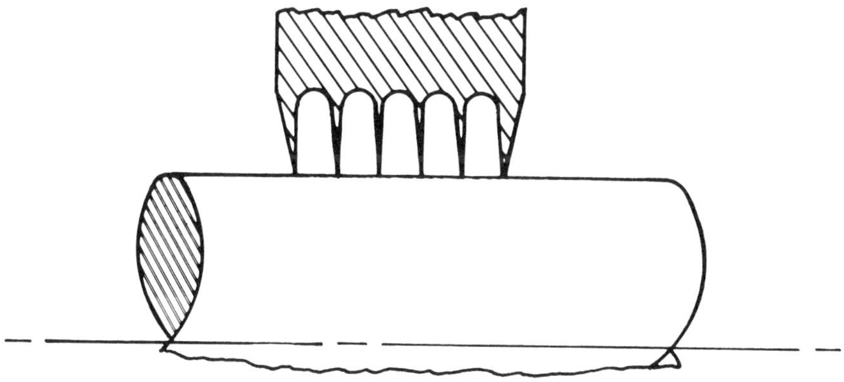

FIG. 2-14 Labyrinth type of noncontacting seal.

2-11-5 Compression Packings

The stuffing box (Fig. 2-13) is used to seal fluids under pressure with either rotating or reciprocating shafts. Sealing between the packing and shaft occurs as a result of axial movement of the gland when the nuts are tightened. Friction between the packing and shaft causes wear, so periodic tightening of the nuts is required.

Packing material is usually obtained in straight lengths of square or rectangular cross section. Pieces are cut off and formed into rings that fit the stuffing box. The choice of packing material depends on the fluid to be sealed. Available packing materials include artificial fibers, asbestos, cotton, graphite, jute, leather, and metals. The metal packings are used for temperature conditions where the other materials are inadequate. The metal packings are formed from foil which is compressed into the proper packing shape.

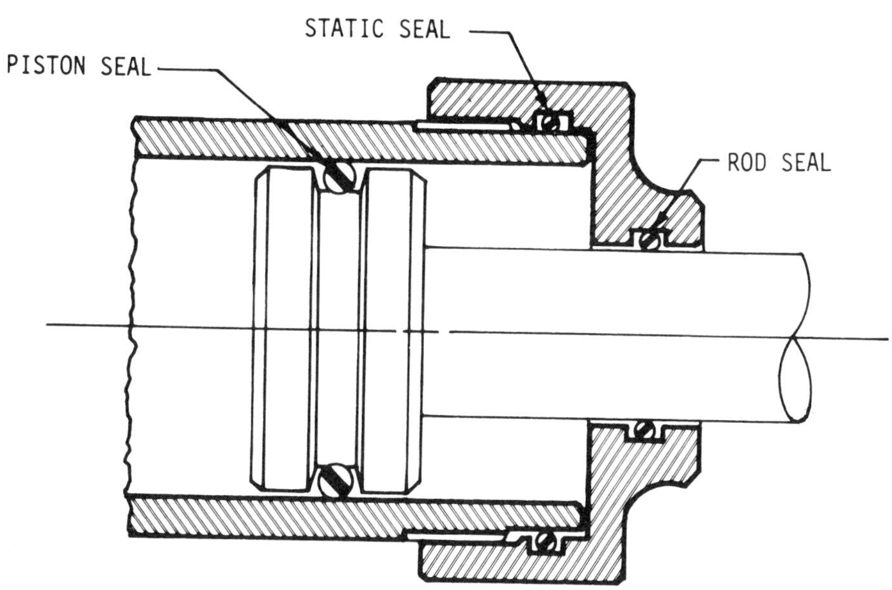

FIG. 2-15 Applications of O-rings as a static seal and as seals for reciprocating motion.

In the design of stuffing boxes, small clearances are provided between the shaft and surrounding parts. The small clearances minimize extrusion of the packing into the clearance spaces.

Valve stems undergo a helical motion rather than a rotary motion when the valve is opened or closed. Investigations into the prevention of valve leakage and wear of valve stems resulted in a procedure for establishing packing dimensions for valve stems [2-15].

2-11-6 Noncontacting Seals

Frictional losses occur with sealing methods that utilize physical contact between a rotating and a stationary part. With high rubbing velocities, friction losses may be a significant factor. Those losses can be eliminated by using a seal that does not require physical contact. A noncontacting seal, however, cannot prevent leakage completely, but does reduce it to a tolerable level.

One method of achieving a low leakage rate is to provide a very small clearance between the shaft and the surrounding housing or bushing. The longer the low-clearance passage, the greater the reduction in leakage.

A type of noncontacting seal called the *labyrinth seal* (Fig. 2-14) is used on such machines as large blowers and steam turbines. It can be used to retain lubricant in the bearings or to seal the working fluid in the machine. Effectiveness of the seal depends on small clearances between the seal and the shaft. In sealing the working fluid, the small clearances create a series of pressure drops between the working fluid and the atmosphere.

Labyrinth seals are usually made from a relatively soft metal such as aluminum or bronze so that the shaft is not damaged if contact between shaft and seal occurs. The simplest type is shown in Fig. 2-14, but other types are also used.

2-12 SEALS FOR RECIPROCATING MOTION

Some of the sealing methods used for rotary motion are also satisfactory for reciprocating motion. O-rings, compression packings, metal sealing rings, and some additional types are used to seal reciprocating rods, shafts, and pistons.

2-12-1 O-Rings

The O-ring is used extensively because of the low installed cost and effectiveness as a seal. It is well adapted to sealing reciprocating motion as well as to use as a static seal. Figure 2-15 shows applications on a piston and piston rod as well as a static seal application. Many such applications are for hydraulic cylinders in which the hydraulic oil acts as the lubricant for the O-rings.

Rectangular-section rings are not suited for reciprocatng motion and are used only in static applications. The shape of the groove for circular-section O-rings for sealing reciprocating motion is the same as for static applications (Fig. 2-8). The recommended groove depth *E,* however, is slightly different for reciprocating motion. Recommended dimensions are available in SAE J120a and in the manufacturer's literature.

An O-ring must seal the clearance space between the reciprocating and stationary

Notes ▪ Drawings ▪ Ideas

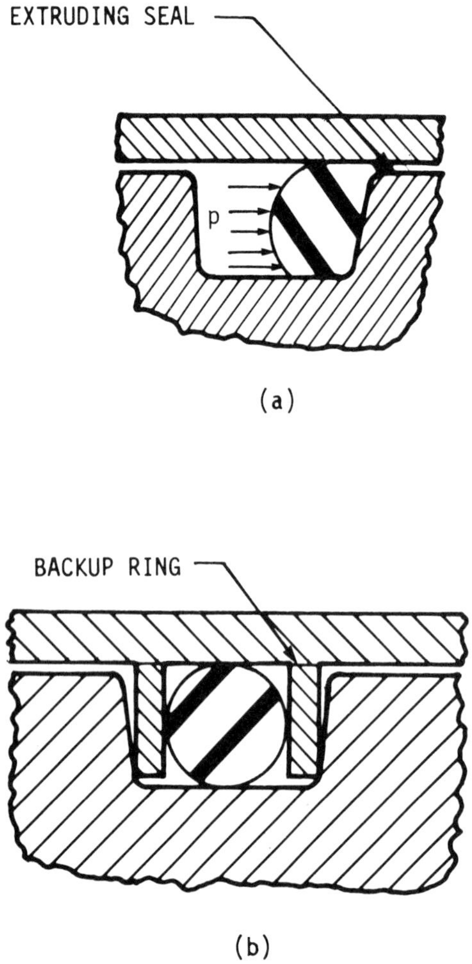

FIG. 2-16 (*a*) Extrusion of O-ring into clearance space due to pressure. (*b*) Use of backup ring to prevent extrusion.

parts, for example, between the piston and cylinder in Fig. 2-15. The amount of clearance that can be permitted depends on the pressure differential across the O-ring and the ring hardness. If the clearance is too great, the O-ring is extruded into the clearance space (Fig. 2-16*a*). The reciprocating motion then tears away small pieces of the O-ring which results in a leaking seal and contamination of the working fluid by the O-ring particles.

The pressure limitations of O-rings can be overcome by the use of backup rings (Fig. 2-16*b*) or other devices that prevent O-ring extrusion. Backup rings are made from leather, plastics, or metal. The metal rings are split like a piston ring for radial compression during assembly into the cylinder.

Recommendations on the combination of clearance and pressure for which backup rings are required vary to some extent among the various O-ring suppliers. Figure 2-17 provides one such recommendation [2-16]. In this figure, if the combination of fluid pressure and maximum gap falls to the right of a hardness curve, backup rings are required. If a piston or rod can be forced to one side of the bore, the maximum gap is the difference between the two diameters. If, however, the radial position of the piston or rod is restrained as by bearings, the radial clearance is the maximum gap.

Both the compression of the O-ring cross section to effect a seal and fluid pressure acting on the seal cause friction forces that oppose reciprocating motion. Information for estimating friction factors is provided in Fig. 2-18 [2-16]. The friction factor due to compression of the cross section can be obtained from Fig. 2-18*a*. The seal compression is expressed as a percentage of the O-ring cross section. The friction factor is multiplied by the circumference of the surface where relative motion occurs to obtain the friction force. For a piston, the circumference of the cylinder bore is used; for a piston rod, the rod circumference applies.

The friction factor for pressure differential across the O-ring is obtained from Fig. 2-18*b*. That factor is multiplied by the projected area of the O-ring to obtain the friction force. For an O-ring in a piston, the projected area is the product of the diameter of the ring cross section and the circumference of the cylinder bore. The total estimated friction force is the sum of the friction forces due to compression and fluid pressure.

The finish of rubbing surfaces should be 8 to 16 μin (0.2 to 0.4 μm) for O-rings with a hardness of 70 Shore A. For rougher surfaces, a higher O-ring hardness should be used to ensure reasonable life.

O-rings may be damaged if they are forced over sharp corners during assembly.

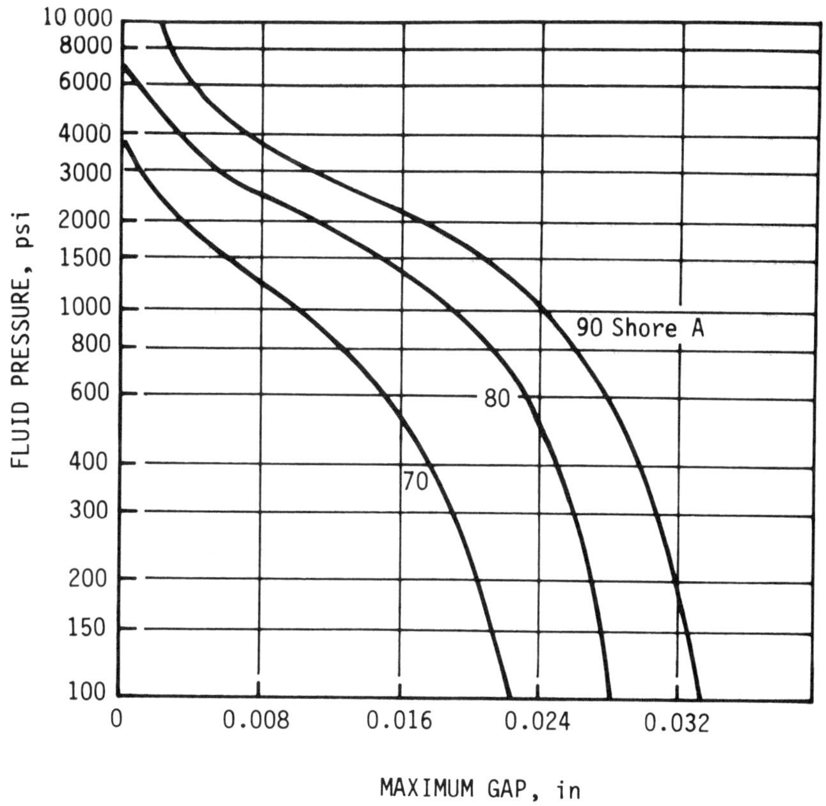

FIG. 2-17 Extrusion limits for O-rings. *(From Ref. [2-16].)*

The addition of chamfers to corners is an inexpensive method of reducing damage. Serious damage to O-rings occurs if they are forced to pass over a hole in a cylinder wall while under pressure. If this occurs, the O-ring expands into the hole and must later be forced back into the groove. This action tends to shear pieces off the ring and thus destroy its ability to seal.

2-12-2 Lip Packings

Cup packings, U-seals, V-ring packings, and other forms of lip packings are used primarily to seal reciprocating motion. The packing material is usually leather, solid rubber, or fabric-reinforced rubber, although other compounds are available for difficult applications. An advantage of leather packings is a low coefficient of friction, on the order of 0.006 to 0.008 depending on the tanning process. Low friction increases the life of a packing because less heat is generated.

Cup packings (Fig. 2-19) were one of the first types of piston seals for hydraulic and pneumatic applications. The fluid pressure expands the cup outward against the cylinder wall and thus seals the piston in the cylinder. This action requires that a double-acting cylinder have two packings in order to seal the pressure in both directions of operation. The inner portion of the piston in Fig. 2-19 is a boss to prevent excessive tightening of the washer against the cup. If the cup is crushed against the piston, good sealing will not be obtained.

Figure 2-20 shows elastomeric U-seals on a double-acting piston. This type of seal is also used on piston rods. They have approximately the same pressure limi-

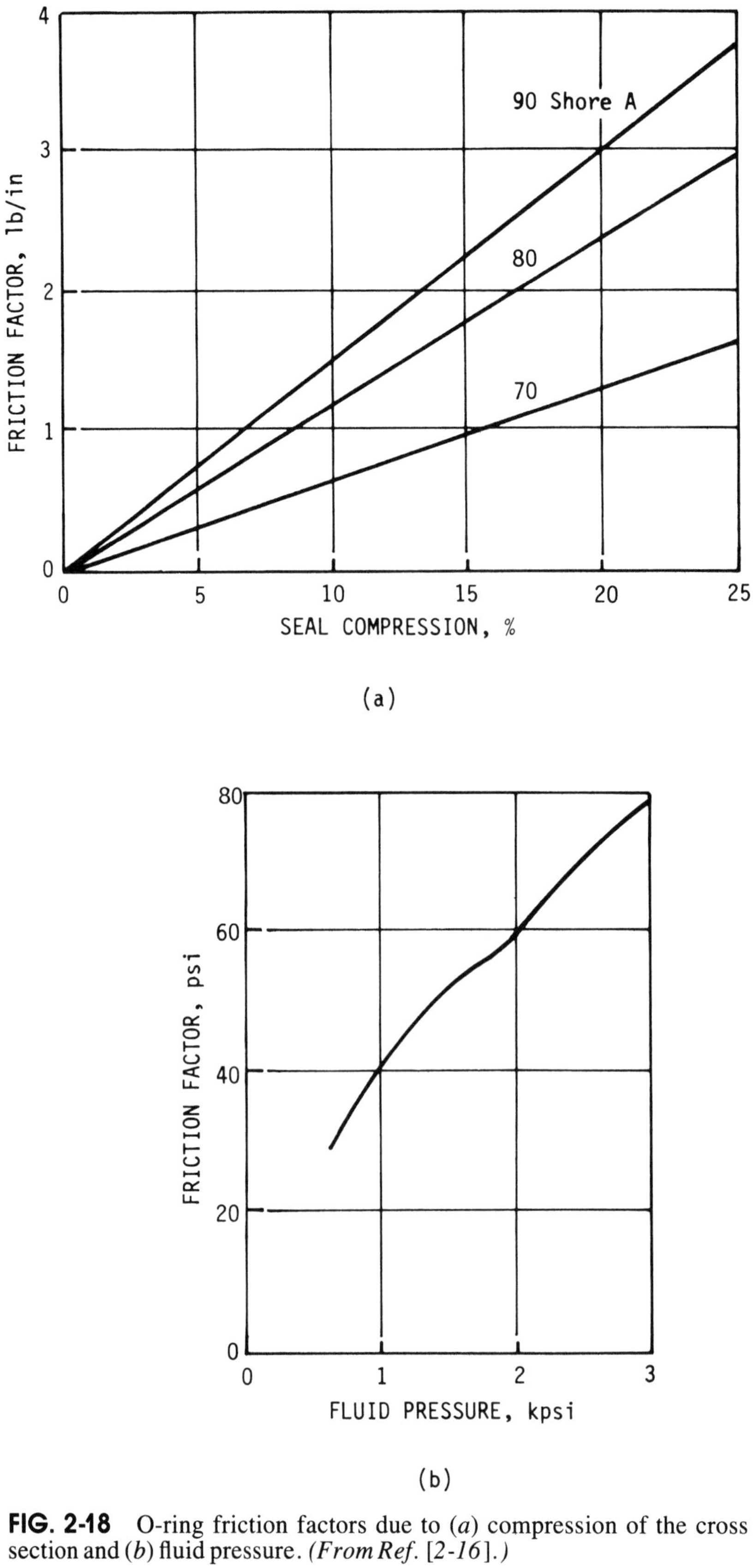

FIG. 2-18 O-ring friction factors due to (*a*) compression of the cross section and (*b*) fluid pressure. *(From Ref.* [2-*16*].*)*

Notes ▪ Drawings ▪ Ideas

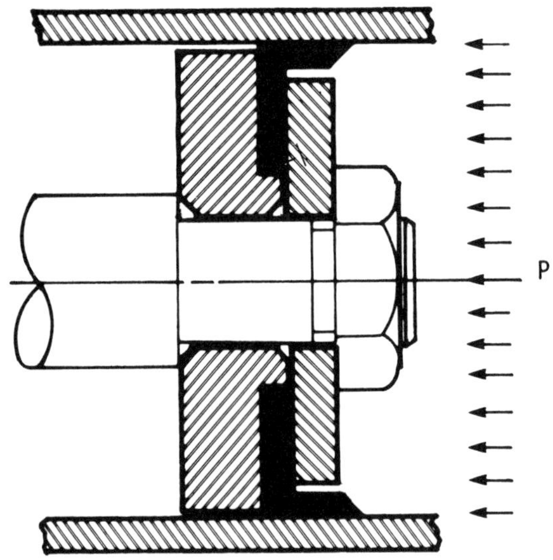

FIG. 2-19 Cup packing for single-acting cylinder.

tations as O-rings, and backup rings are required for higher pressures. When a U-seal is made of leather, a filler is required between the lips to prevent collapse of the seal.

Rod scrapers are used on piston rods of hydraulic cylinders that are exposed to harsh environments. The purpose is to exclude mud, dust, and ice from the cylinder. A typical rod scraper is composed of a polyurethane element bonded to a metal shell which is pressed into the end cap of the cylinder. Molybdenum disulfide is sometimes added to the polyurethane to reduce friction. A rod scraper added to the end cap of a cylinder is shown in Fig. 2-21. The sealing lip is pointed outward to remove foreign material when the piston rod is retracted.

The chief use of the V-ring packing (Fig. 2-22) is for sealing piston rods or reciprocating shafts, although it can also be used to seal pistons. The ability to seal fluids

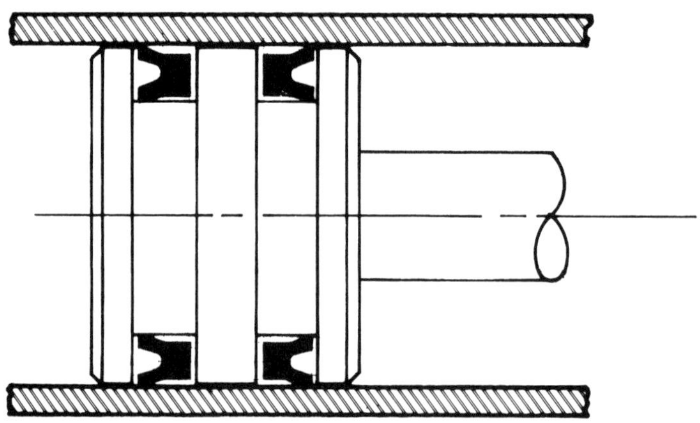

FIG. 2-20 U-seals for a double-acting piston.

Notes ▪ Drawings ▪ Ideas

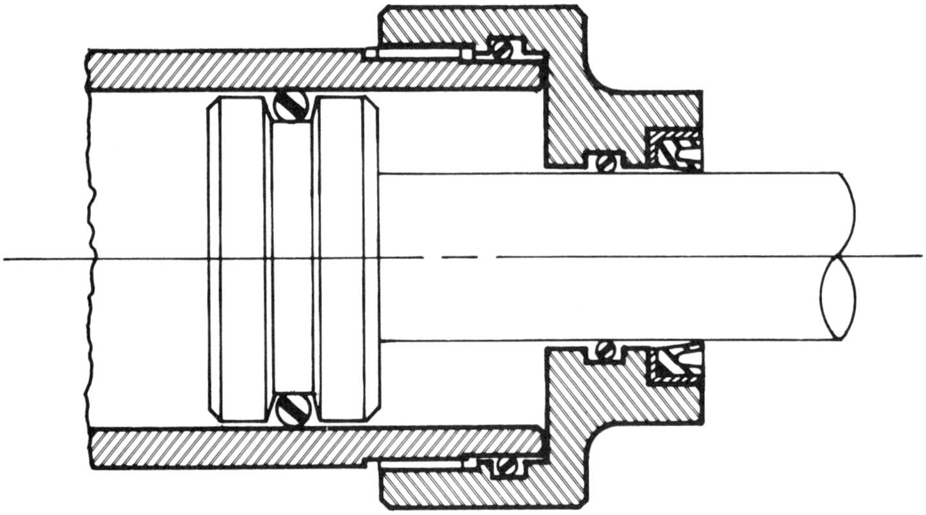

FIG. 2-21 Rod scraper on piston rod.

under pressure depends on the type of packing material and number of packings used. The V-ring packing is considered superior to other lip types for sealing high pressures, especially above 50 000 psi (345 N/mm^2).

The V shape of the packing is obtained through the adapters that support the rings. Fluid pressure then expands the packing against the shaft and housing to seal the fluid. A continuous packing provides better sealing than a series of split rings, although the latter are easier to install and remove.

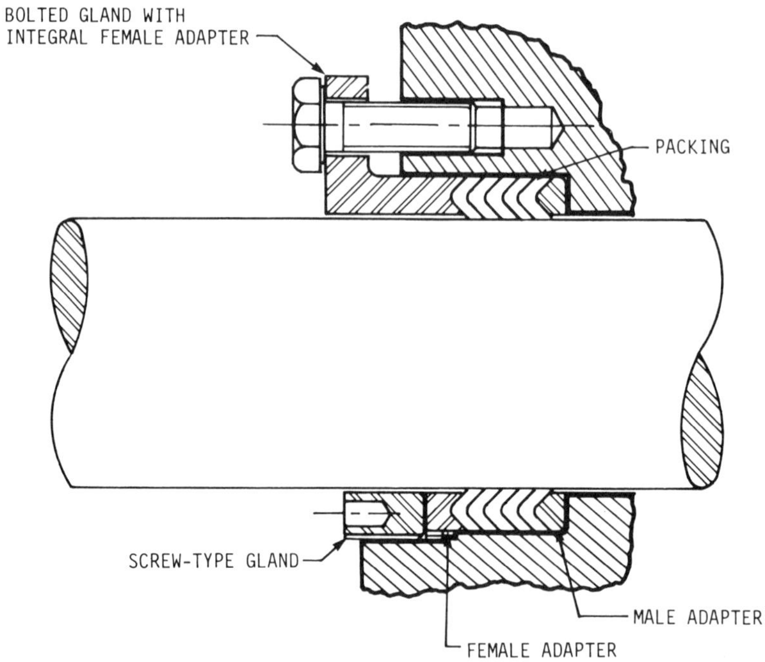

FIG. 2-22 V-ring packing for a reciprocating shaft.

2-12-3 Piston Rings

Piston rings for automotive engine applications are made of gray cast iron. The rings are used to seal gases in the cylinder and to restrict oil to the crankcase. Standard dimensions for piston rings and grooves are provided in SAE J929a and J1201.

Piston rings must be split for assembly over the piston. This requires shaping the ring so that it will provide a uniform radial force against the cylinder wall. Piston ring manufacturers have developed methods of attaining this objective.

REFERENCES

2-1 *Standard Classification System for Nonmetallic Gasket Materials* ANSI/ ASTM F104-79a, American Society for Testing and Materials.

2-2 D. E. Czernik, J. C. Moerk, Jr., and F. A. Robbins, ''The Relationship of a Gasket's Physical Properties to the Sealing Phenomena,'' SAE paper 650431, May 1965.

2-3 D. E. Czernik, ''Recent Developments and New Approaches in Mechanical and Chemical Gasketing,'' SAE paper 810367, Feb. 1981.

2-4 V. M. Faires, *Design of Machine Elements,* MacMillan, 1955.

2-5 D. J. McDowell, ''Choose the Right Gasket Material,'' *Assembly Engineering,* Oct. 1978.

2-6 H. A. Rothbart, *Mechanical Design and Systems Handbook,* 2nd ed., McGraw-Hill, 1985, Sec. 27.4.

2-7 *Armstrong Gasket Design Manual,* Armstrong Cork Co., Lancaster, Pa., 1978.

2-8 J. W. Oren, ''Creating Gasket Seals with Rigid Flanges,'' SAE paper 810362, Feb. 1981.

2-9 D. E. Czernik, ''Gasketing the Internal Combustion Engine,'' SAE paper 800073, Feb. 1980.

2-10 The American Society of Mechanical Engineers, Code for Pressure Vessels, Sec. VIII, Div. 1, App. 2, 1980.

2-11 J. J. Whalen, ''How to Select the Right Gasket Material,'' *Product Engineering,* Oct. 1960.

2-12 D. E. Czernik, ''Sealing Today's Engines,'' *Fleet Maintenance and Specifying,* Irving-Cloud, July 1977.

2-13 Leonard J. Martini, ''Sealing Rotary Shafts with O-Rings,'' *Machine Design,* May 26, 1977, pp. 97–99.

2-14 Bert Robins, ''Radial Lip Seals—Are Two Too Many?'' *Power Transmission Design,* Oct. 1982, pp. 73, 74.

2-15 L. I. Ezekoye and J. A. George, ''Valve Packings That Don't Leak,'' *Machine Design,* Jan. 20, 1977, pp. 142, 143.

2-16 Wes J. Ratelle, ''Seal Selection: Beyond Standard Practice,'' *Machine Design,* Jan. 20, 1977, pp. 133–137.

chapter **3**
ROLLING-CONTACT BEARINGS

CHARLES R. MISCHKE, Ph.D., P.E.
Professor of Mechanical Engineering
Iowa State University
Ames, Iowa

GLOSSARY OF SYMBOLS

a	Exponents; $a = 3$ for ball bearings; $a = \frac{10}{3}$ for roller bearings
AF	Application factor
b	Weibull shape parameter
C_s	Static load rating
C_{10}	Basic load rating or basic dynamic load rating
f	Fraction
F	Load
F_a	Axial load
F_{eq}	Equivalent radial load
F_i	ith equivalent radial load
F_r	Radial load
I	Integral
L	Life measure, r or h
L_D	Desired or design life measure
L_R	Rating life measure
L_{10}	Life measure exceeded by 90 percent of bearings tested
n	Design factor
n_D	Desired or design, rpm
n_i	Application or design factor at ith level
n_R	Rating, r/min
R	Reliability
V	Rotation factor; inner ring rotations, $V = 1$; outer ring, $V = 1.20$
x	Life measure in Weibull survival equation
x_0	Weibull guaranteed life parameter
X	Radial factor for equivalent load prediction
Y	Thrust factor for equivalent load prediction

θ Weibull characteristic life parameter, rotation angle

ϕ Period of cyclic variation, rad

3-1 INTRODUCTION

Figures 3-1 to 3-12 illustrate something of the terminology and the wide variety of rolling-contact bearings available to the designer. Catalogs and engineering manuals can be obtained from bearing manufacturers, and these are very comprehensive and of excellent quality. In addition, most manufacturers are anxious to advise designers on specific applications. For this reason the material in this chapter is concerned mostly with providing the designer an independent viewpoint.

Rolling-contact bearings use balls and rollers to exploit the small coefficients of friction when hard bodies roll on each other. Balls and rollers are kept separated and equally spaced by a separator (cage, or retainer). This device, which is essential to proper bearing functioning, is responsible for additional friction. Table 3-1 (page 100) gives friction coefficients for several types of bearings [3-1]. Consult a manufacturer's catalog for equations for estimating friction torque as a function of bearing mean diameter, load, basic load rating, and lubrication detail. See also Chap. 1.

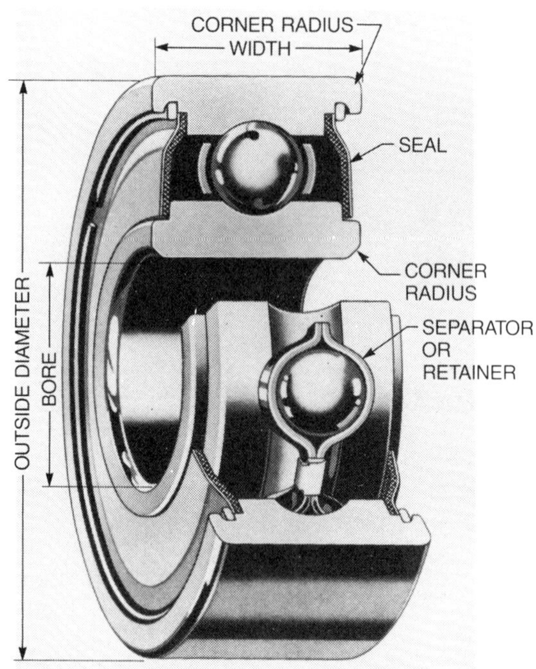

FIG. 3-1 Photograph of a deep-groove precision ball bearing with metal two-piece cage and dual seals to illustrate rolling-bearing terminology. *(The Barden Corporation.)*

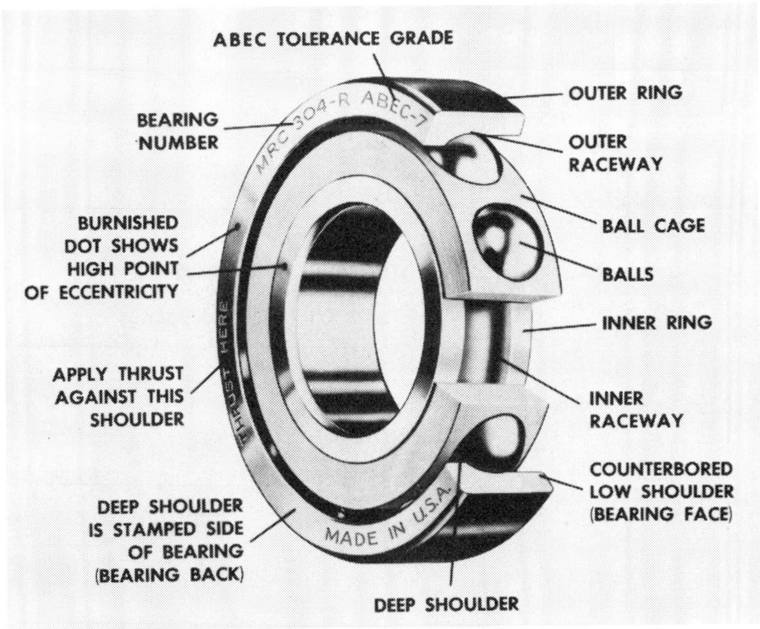

FIG. 3-2 Photograph of a precision ball bearing of the type generally used in machine-tool applications to illustrate terminology. *(Bearings Division, TRW Industrial Products Group.)*

FIG. 3-3 Rolling bearing with spherical rolling elements to permit misalignment up to ±3° with an unsealed design. The sealed bearing, shown above, permits misalignment to ±2°. *(McGill Manufacturing Company, Inc.)*

FIG. 3-4 A heavy-duty cage-guided needle roller bearing with machined race. Note the absence of an inner ring, but standard inner rings can be obtained. *(McGill Manufacturing Company, Inc.)*

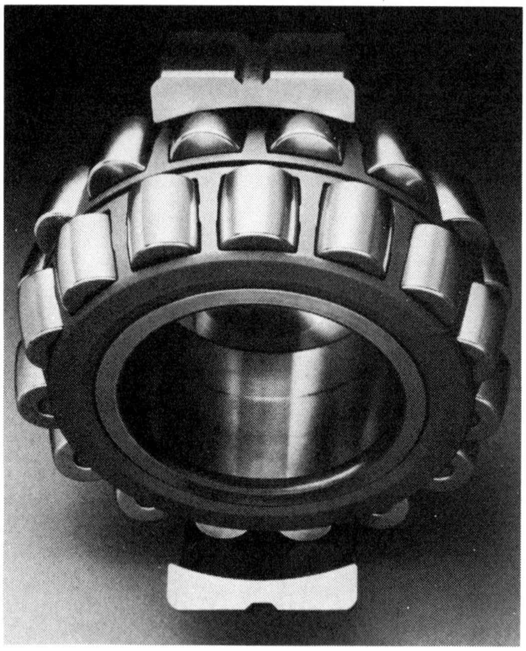

FIG. 3-5 A spherical roller bearing with two rows of rollers running on a common sphered raceway. These bearings are self-aligning to permit misalignment resulting from either mounting or shaft deflection under load. *(SKF Industries, Inc.)*

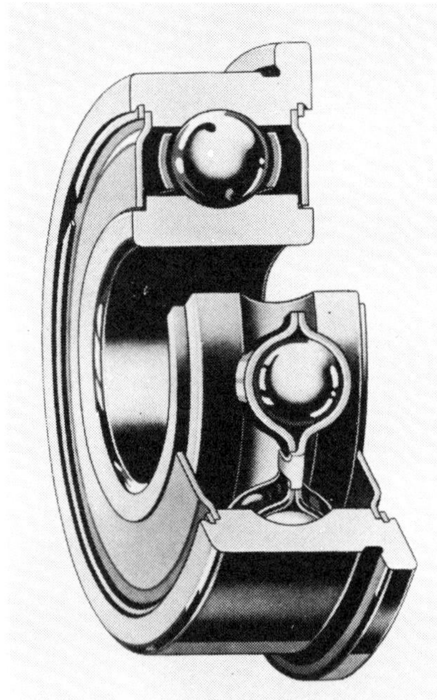

FIG. 3-6 Shielded, flanged, deep-groove ball-bearing. Shields serve as dirt barriers; flange facilitates mounting the bearing in a through-bored hole. *(The Barden Corporation.)*

FIG. 3-7 Ball thrust bearing. *(The Torrington Company.)*

FIG. 3-8 Spherical roller thrust bearing. *(The Torrington Company.)*

Notes ▪ Drawings ▪ Ideas

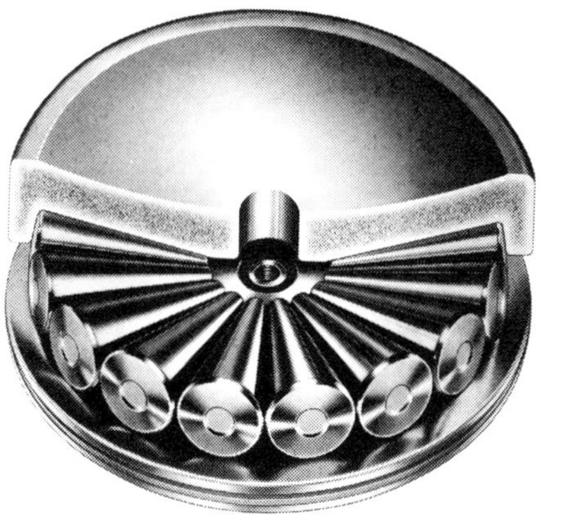

FIG. 3-9 Tapered-roller thrust bearing. *(The Torrington Company.)*

FIG. 3-10 Tapered-roller bearing; for axial loads, thrust loads, or combined axial and thrust loads. *(The Timken Company.)*

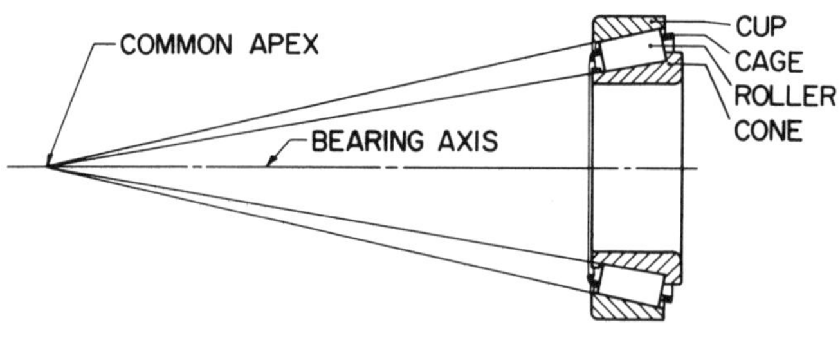

FIG. 3-11 Basic principle of a tapered-roller bearing with nomenclature. *(The Timken Company.)*

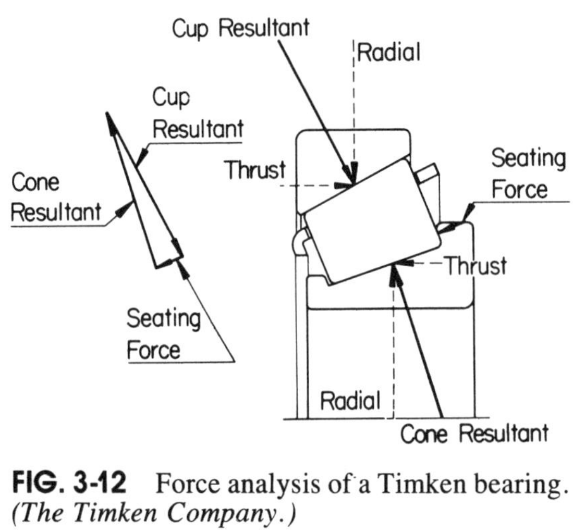

FIG. 3-12 Force analysis of a Timken bearing. *(The Timken Company.)*

Notes ▪ Drawings ▪ Ideas

TABLE 3-1 Coefficients of Friction

Bearing type	Coefficient of friction μ
Self-aligning ball	0.0010
Cylindrical roller with flange-guided short rollers	0.0011
Ball thrust	0.0013
Single-row ball	0.0015
Spherical roller	0.0018
Tapered roller	0.0018

SOURCE: Ref. [3-1].

Permissible speeds are influenced by bearing size, properties, lubrication detail, and operating temperatures. The speed varies inversely with mean bearing diameter. For additional details consult any manufacturer's catalog.

Some of the guidelines for selecting bearings, which are valid more often than not, are as follows:

- Ball bearings are the less expensive choice in the smaller sizes and under lighter loads, while roller bearings are less expensive for larger sizes and heavier loads.
- Roller bearings are more satisfactory under shock or impact loading than ball bearings.
- Ball-thrust bearings are for pure thrust loading only. At high speeds a deep-groove or angular-contact ball bearing usually will be a better choice, even for pure thrust loads.
- Self-aligning ball bearings and cylindrical roller bearings have very low friction coefficients.
- Deep-groove ball bearings are available with seals built into the bearing so that the bearing can be prelubricated to operate for long periods without attention.
- Although rolling-contact bearings are "standardized" and easily selected from vendor catalogs, there are instances of cooperative development by customer and vendor involving special materials, hollow elements, distorted raceways, and novel applications. Consult your bearing specialist.

It is possible to obtain an estimate of the basic static load rating C_s. For ball bearings,

$$C_s = Mn_b d_b^2 \tag{3-1}$$

For roller bearings

$$C_s = Mn_r l_e d \tag{3-2}$$

where C_s = basic static loading rating, pounds (lb) [kilonewtons (kN)]
n_b = number of balls
n_r = number of rollers
d_b = ball diameter, inches (in) [millimeters (mm)]
d = roller diameter, in (mm)
l_e = length of single-roller contact line, in (mm)

Values of the constant M are listed in Table 3-2.

Notes ▪ Drawings ▪ Ideas

TABLE 3-2 Value of Constant M for Use in Eq. (3-1) and (3-2)

Type of bearing	Constant M	
	C_s, lb	C_s, kN
Radial ball	1.78×10^3	5.11×10^3
Ball thrust	7.10×10^3	20.4×10^3
Radial roller	3.13×10^3	8.99×10^3
Roller thrust	14.2×10^3	40.7×10^3

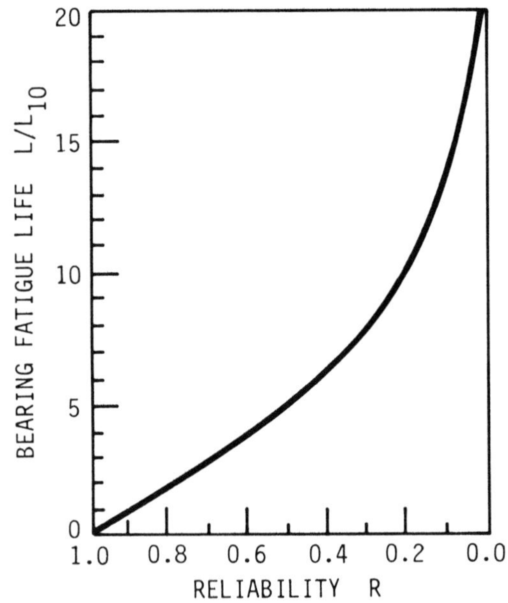

FIG. 3-13 Survival function representing endurance tests on rolling-contact bearings from data accumulated by SKF Industries, Inc. *(From Ref.* [*3-2*].*)*

3-2 LOAD-LIFE RELATION FOR CONSTANT RELIABILITY

When proper attention is paid to a rolling-contact bearing so that fatigue of the material is the only cause of failure, then nominally identical bearings exhibit a reliability-life-measure curve, as depicted in Fig. 3-13. The *rating life* is defined as the life measure (revolutions, hours, etc.) which 90 percent of the bearings will equal or exceed. This is also called the L_{10} life or the B_{10} life. When the radial load is adjusted so that the L_{10} life is 1 000 000 revolutions (r), that load is called the *basic load rating* C (SKF Industries, Inc.). The Timken Company rates its bearings at 90 000 000. Whatever the rating basis, the life L can be normalized by dividing by the rating life L_{10}. The *median life* is the life measure equaled or exceeded by half of the bearings. Median life is roughly 5 times rating life.

For steady radial loading, the life at which the first tangible evidence of surface fatigue occurs can be predicted from

$$F^a L = \text{constant} \tag{3-3}$$

where $a = 3$ for ball bearings and $a = \frac{10}{3}$ for cylindrical and tapered-roller bearings. At constant reliability the load and life at condition 1 can be related to the load and life at condition 2 by Eq. (3-3). Thus

$$F_1^a L_1 = F_2^a L_2 \tag{3-4}$$

If F_1 is the basic load rating C_{10}, then L_1 is the rating life L_{10}, so

$$C_{10} = \left(\frac{L}{L_{10}}\right)^{1/a} (F) \tag{3-5}$$

Notes ▪ Drawings ▪ Ideas

If L_R in hours and n_R is in revolutions per minute, then $L_{10} = 60L_Rn_R$. It follows that

$$C_{10} = F_D\left(\frac{L_Dn_D}{L_Rn_R}\right)^{1/a} \tag{3-6}$$

where the subscript D refers to desired (or design) and the subscript R refers to rating conditions.

3-3 SURVIVAL RELATION AT STEADY LOAD

Figure 3-14 shows how reliability varies as the loading is modified [3-2]. Equation (3-5) allows the ordinate to be expressed as either F/C_{10} or L/L_{10}. Figure 3-14 is based on more than 2500 SKF bearings. If Figs. 3-13 and 3-14 are scaled for recovery of coordinates, then the reliability can be tabulated together with L/L_{10}. Machinery applications use reliabilities exceeding 0.94. An excellent curve fit can be realized by using the three-parameter Weibull distribution. For this distribution the reliability can be expressed as

$$R = \exp\left[-\left(\frac{x - x_0}{\theta - x_0}\right)^b\right] \tag{3-7}$$

where x = life measure, x_0 = Weibull guaranteed life measure, θ = Weibull characteristic life measure, and b = Weibull shape factor. Using the 18 points in Table

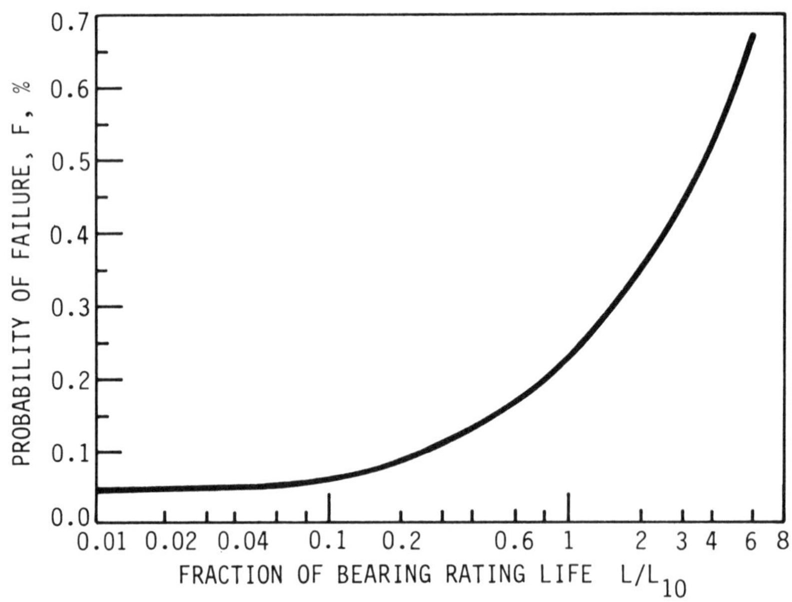

FIG. 3-14 Survival function at higher reliabilities based on more than 2500 endurance tests by SKF Industries, Inc. *(From Ref. [3-2].)* The three-parameter Weibull constants are $\theta = 4.459$, $b = 1.483$, and $x_0 = 0.02$ when $x = L/L_{10} = Ln/(L_Rn_R)$.

TABLE 3-3 Survival Equation Points at Higher Reliabilities*

Reliability R	Life measure L/L_{10}	Reliability R	Life measure L/L_{10}
0.94	0.67	0.994	0.17
0.95	0.60	0.995	0.15
0.96	0.52	0.996	0.13
0.97	0.435	0.997	0.11
0.975	0.395	0.9975	0.095
0.98	0.35	0.998	0.08
0.985	0.29	0.9985	0.07
0.99	0.23	0.999	0.06
0.992	0.20	0.9995	0.05

*Scaled from Ref. [3-2], Fig. 2.

3-3 with $x_0 = 0.02$, $\theta = 4.459$, and $b = 1.483$, we see that Eq. (3-7) can be particularized as

$$R = \exp\left[-\left(\frac{L/L_{10} - 0.02}{4.439}\right)^{1.483}\right] \tag{3-8}$$

For example, for $L/L_{10} = 0.1$, Eq. (3-8) predicts $R = 0.9974$.

3-4 RELATING LOAD, LIFE, AND RELIABILITY GOAL

If Eq. (3-3) is plotted on log-log coordinates, Fig. 3-15 results. The FL loci are rectified, while the parallel loci exhibit different reliabilities. The coordinates of point A are the rating life and the basic load rating. Point D represents the desired (or design) life and the corresponding load. A common problem is to select a bearing

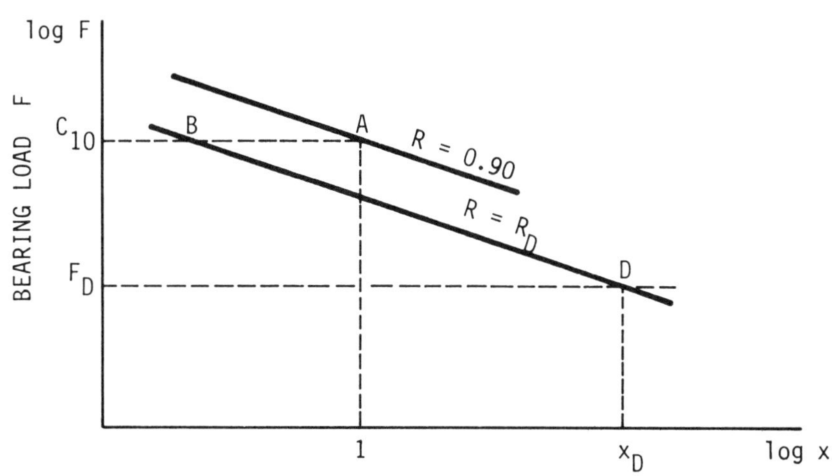

FIG. 3-15 Reliability contours on a load-life plot useful for relating catalog entry, point A, to design goal, point D.

which will provide a life L_D while carrying load F_D and exhibit a reliability R_D. Along line BD, constant reliability prevails, and Eq. (3-4) applies:

$$F_B = F_D\left(\frac{x_D}{x_B}\right)^{1/a} \tag{3-9}$$

Along line AB the reliability changes, but the load is constant and Eq. (3-7) applies. Thus

$$R = \exp\left[-\left(\frac{x - x_0}{\theta - x_0}\right)^b\right] \tag{3-10}$$

Now solve this equation for x and particularize it for point B, noting that $R_D = R_B$.

$$x_B = x_0 + (\theta - x_0)\left(\ln\frac{1}{R_D}\right)^{1/b} \tag{3-11}$$

Substituting Eq. (3-11) into Eq. (3-9) yields

$$F_B = C_{10} = F_D\left\{\frac{x_D}{x_0 + (\theta - x_0)[\ln(1/R_D)]^{1/b}}\right\}^{1/a} \tag{3-12}$$

For reliabilities greater than 0.90, which is the usual case, $\ln(1/R) \cong 1 - R$ and Eq. (3-12) simplifies as follows:

$$C_{10} = F_D\left[\frac{x_D}{x_0 + (\theta - x_0)(1 - R)^{1/b}}\right]^{1/a} \tag{3-13}$$

The desired life measure x_D can be expressed most conveniently in millions of revolutions (for SKF).

EXAMPLE 1. If a ball bearing must carry a load of 800 lb for 50×10^6 and exhibit a reliability of 0.99, then the basic load rating should equal or exceed

$$\begin{aligned} C_{10} &= 800\left[\frac{50}{0.02 + (4.439)(1 - 0.99)^{1/1.483}}\right]^{1/3} \\ &= 4890 \text{ lb} \end{aligned}$$

This is the same as 21.80 kN, which corresponds to the capability of a 02 series 40-mm-bore ball bearing. Since selected bearings have different basic load ratings from those required, a solution to Eq. (3-13) for reliability extant after specification is useful:

$$R = 1 - \left[\frac{x_D - x_0(C_{10}/F_D)^a}{(\theta - x_0)(C_{10}/F_D)^a}\right]^b \tag{3-14}$$

////

EXAMPLE 2. If the bearing selected for Example 1, a 02 series 50-mm bore, has a basic load rating of 26.9 kN, what is the expected reliability? And $C_{10} = 26.9(10^3)/445 = 6045$ lb. So

$$R = 1 - \left[\frac{50 - 0.02(6045/800)^3}{(4.439)(6045/800)^3}\right]^{1.483} = 0.9966$$

////

Notes ▪ Drawings ▪ Ideas

The previous equations can be adjusted to a two-parameter Weibull survival equation by setting x_0 to zero and using appropriate values of θ and b. For bearings rated at a particular speed and time, substitute $L_D n_D/(L_R n_R)$ for x_D.

The survival relationship for Timken tapered-roller bearings is shown graphically in Fig. 3-16, and points scaled from this curve form the basis for Table 3-4. The survival equation turns out to be the two-parameter Weibull relation:

$$R = \exp\left[-\left(\frac{x}{\theta}\right)^b\right] = \exp\left[-\left(\frac{L/L_{10}}{4.890}\right)^{1.4335}\right] \tag{3-15}$$

The equation corresponding to Eq. (3-13) is

$$\begin{aligned} C_{10} &= F_D\left[\frac{x_D}{\theta(1-R)^{1/b}}\right]^{1/a} \\ &= F_D\left(\frac{x_D}{\theta}\right)^{1/a}(1-R)^{-1/ab} \end{aligned} \tag{3-16}$$

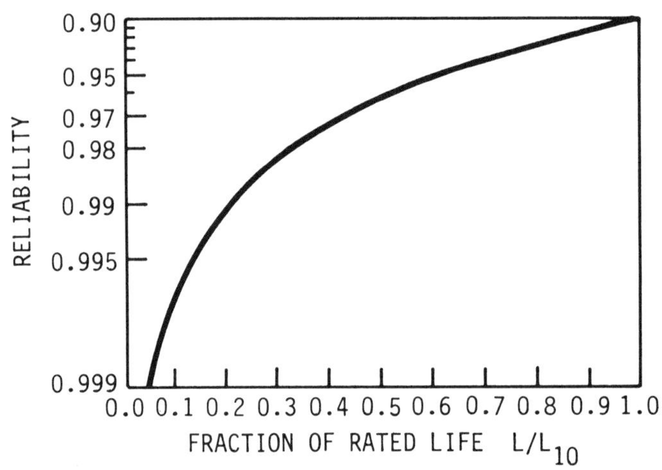

FIG. 3-16 Survival function at higher reliabilities based on the Timken Company tapered-roller bearings. The curve fit is a two-parameter Weibull function with constants $\theta = 4.890$ and $b = 1.4335$ ($x_0 = 0$) when $x = Ln/(L_R n_R)$. Timken uses $\theta = 4.48$, $b = 3/2$. *(From Ref. [3-3].)*

TABLE 3-4 Survival Equation Points for Tapered-Roller Bearings*

Reliability R	Life measure L/L_{10}	Reliability R	Life measure L/L_{10}
0.90	1.00	0.96	0.53
0.91	0.92	0.97	0.43
0.92	0.86	0.98	0.325
0.93	0.78	0.99	0.20
0.94	0.70	0.995	0.13
0.95	0.62	0.999	0.04

*Scaled from Ref. [3-3], Fig. 4.

Notes ▪ Drawings ▪ Ideas

And the equation corresponding to Eq. (3-14) is

$$R = 1 - \left(\frac{x_D}{\theta}\right)^b \left(\frac{C_{10}}{F_D}\right)^{-ab} \tag{3-17}$$

EXAMPLE 3. A Timken tapered-roller bearing is to be selected to carry a radial load of 4 kN and have a reliability of 0.99 at 1200 hours (h) and a speed of 600 revolutions per minute (rpm). Thus

$$x_D = \frac{L_D n_D}{L_R n_R} = \frac{1200(600)}{3000(500)} = 0.480$$

and

$$C_{10} = 4\left[\frac{0.48}{4.48(1 - 0.99)^{1/1.5}}\right]^{3/10} = 5141 \text{ N}$$

Timken bearings are rated in U.S. Customary System (USCS) units or in newtons (N); therefore, a basic load rating of 5141 N or higher is to be sought.

For any bearings to be specified, check with the manufacturer's engineering manual for survival equation information. This is usually in the form of graphs, nomograms, or equations. ////

3-5 COMBINED RADIAL AND THRUST LOADINGS

Ball bearings can resist some thrust loading simultaneously with a radial load. The equivalent radial load is the constant pure radial load which inflicts the same damage to the bearing per revolution as the combination. A common form for weighting the radial load F_r and the axial load F_a is

$$F_e = VXF_r + YF_a \tag{3-18}$$

where F_e = equivalent radial load. The weighting factors X and Y are given for each bearing type in the manufacturer's engineering manual. The parameter V distinguishes between inner-ring rotation $V = 1$, and outer-ring rotation, $V = 1.20$. A common form of Eq. (3-18) is

$$F_e = \max(VF_r, X_1 VF_r + Y_1 F_a, X_2 VF_r + Y_2 F_a, \ldots) \tag{3-19}$$

3-6 APPLICATION FACTORS

In machinery applications the peak radial loads on a bearing are different from the nominal or average load owing to a variation in torque or other influences. For a number of situations in which there is a body of measurement and experience, bearing manufacturers tabulate application factors that are used to multiply the average load to properly account for the additional fatigue damage resulting from the fluctuations. Such factors perform the same function as a design factor. In previous equations F_D is replaced by nF_D, or AF(F_D),where AF is the application factor.

Notes ▪ Drawings ▪ Ideas

3-7 VARIABLE LOADING

At constant reliability the current F^aL product measures progress toward failure. The area under the F^a versus L curve at failure is an index to total damage resulting in failure. The area under the F^aL locus at any time prior to failure is an index to damage so far. If the radial load or equivalent radial load varies during a revolution or several revolutions in a periodic fashion, then the equivalent radial load is related to the instantaneous radial load by

$$F_{eq} = \left(\frac{1}{\phi} \int_0^{\phi} F^a \, d\theta \right)^{1/a} \tag{3-20}$$

where ϕ = period of the variation—2π for repetition every revolution, 4π for repetition every second revolution, etc. (see Fig. 3-17).

EXAMPLE 4. A bearing load is given by $F(\theta) = 1000 \sin \theta$ in pound force. Estimate the equivalent load by using Simpson's rule,

$$F_{eq} = \left[\frac{1}{\pi} \int_0^{\pi} (1000 \sin \theta)^{10/3} \, d\theta \right]^{3/10} = 762 \text{ lb}$$

When equivalent loads are applied in a stepwise fashion, the equivalent radial load is expressible by

$$F_{eq} = \left[\sum_{i=1}^{k} f_i (n_i F_i)^a \right]^{1/a} \tag{3-21}$$

where f_i = fraction of revolution at load F_i
n_i = application or design factor
F_i = ith equivalent radial load
a = applicable exponent—3 for ball bearings and $\frac{10}{3}$ for roller bearings

EXAMPLE 5. A four-step loading cycle is applied to a ball bearing. For one-tenth of the time, the speed is 1000 rpm, F_r = 800 lb, and F_a = 400 lb; for two-tenths of the time the speed is 1200 rpm, F_r = 1000 lb, and F_a = 500 lb; for three-tenths of

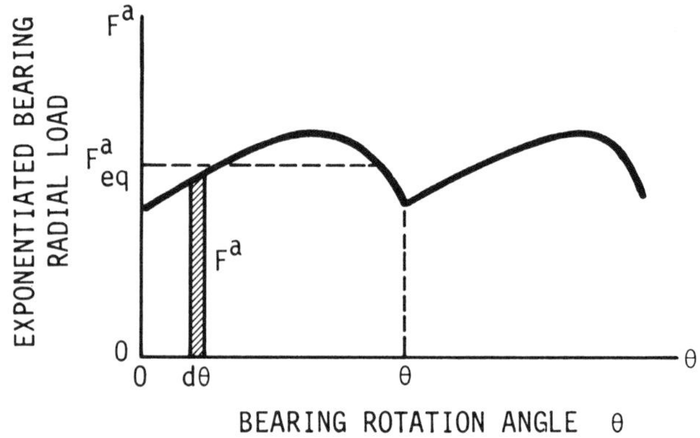

FIG. 3-17 Equivalent radial load when load varies periodically with angular position.

the time the speed is 1500 rpm, $F_r = 1500$ lb, and $F_a = 700$ lb; for four-tenths of the time the speed is 800 rpm, $F_r = 1100$ lb, and $F_a = 500$ lb. For this shallow-angle, angular-contact ball bearing $X_1 = 1$, $Y_1 = 1.25$, $X_2 = 0.45$, $Y_2 = 1.2$, and $V = 1$. This loading cycle is also depicted in Fig. 3-18.

The first step in the solution is to create Table 3-5. The equivalent radial load is

$$F_{eq} = [0.090(1430)^3 + 0.216(2031)^3 + 0.405(2969)^3 + 0.288(2588)^3]^{1/3} = 2604 \text{ lb}$$

Without the use of design factors, the equivalent radial load is

$$F_{eq} = [0.090(1300)^3 + 0.216(1625)^3 + 0.405(2375)^3 + 0.288(1725)^3]^{1/3} = 2002 \text{ lb}$$

The overall design factor is 2604/2002, or 1.30. If this sequence were common in a machinery application, a bearing manufacturer might recommend an application factor of 1.30 for this particular application. ////

3-8 MISALIGNMENT

The inner ring of a rolling-contact bearing is tightly fitted to the shaft, and the axis of rotation is oriented, as is the shaft centerline. The outer ring is held by some form of housing, and its axis is oriented as demanded by the housing. As the shaft deflects under load, these two axes lie at an angle to each other. This misalignment for very small angles is accommodated in "slack," and no adverse life consequences are exhibited. As soon as the slack is exhausted, the intended deflection is resisted and the bearing experiences unintended loading. Life is reduced below prediction levels. A shaft design which is too limber does not fail, but bearings are replaced with much greater frequency. It is too easy to be critical of bearings when the problem lies in the shaft design.

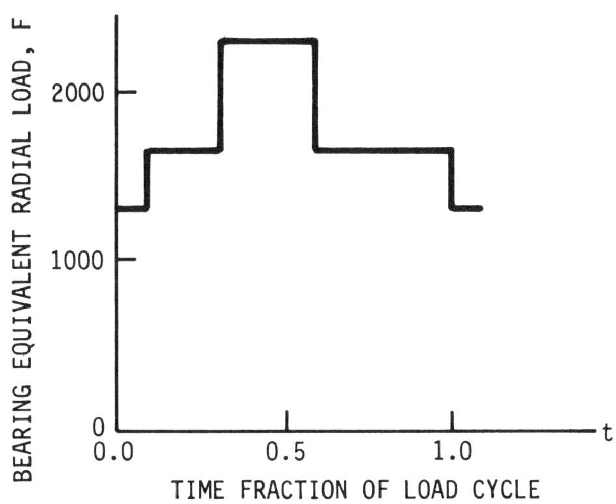

FIG. 3-18 Loading cycle: one-tenth of time at 1000 rpm, $F_r = 800$, $F_a = 400$; two-tenths of time at 1200 rpm, $F_r = 1000$, $F_a = 500$; three-tenths of time at 1500 rpm, $F_r = 1500$, $F_a = 700$; four-tenths of time at 800 rpm, $F_r = 1100$, $F_a = 500$; $X_1 = 1$, $Y_1 = 1.25$, $X_2 = 0.45$, $Y_2 = 1.2$, $V = 1$.

TABLE 3-5 Tabulation for Example 5

Time fraction	Speed, rpm	Product	Revolution fraction f_i	Radial load F_r	Axial load F_a	Equivalent load F_e	Application factor AF	Product (AF)(F_e)
0.1	1000	100	0.090	800	400	1300	1.1	1430
0.2	1200	240	0.216	1000	500	1625	1.25	2031
0.3	1500	450	0.405	1500	700	2375	1.25	2969
0.4	800	320	0.288	1100	500	1725	1.50	2588
		1110						

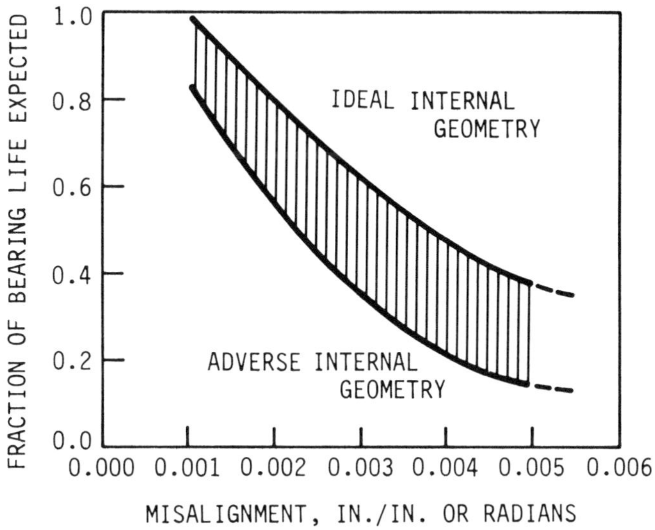

FIG. 3-19 Fractional bearing life to be expected as a function of misalignment in line-contact bearings. *(From Ref. [3-4], Fig. 11.)*

Figure 3-19 shows the dramatic fractional life reduction owing to misalignment in line-contact bearings [3-4].

If there is misalignment, it should not exceed 0.001 radian (rad) in cylindrical and tapered-roller bearings, 0.0087 rad for spherical ball bearings, or about 0.004 rad for deep-groove ball bearings. Self-aligning ball or spherical roller bearings are more tolerant of misalignment. The bibliography of Ref. [3-4] is extensive on this subject.

REFERENCES

3-1 *SKF Engineering Data,* SKF Industries, Inc., Philadelphia, 1979.

3-2 T. A. Harris, "Predicting Bearing Reliability," *Machine Design,* vol. 35, no. 1, Jan. 3, 1963, pp. 129–132.

3-3 *Bearing Selection Handbook,* The Timken Company, Canton, Ohio, revised 1986.

3-4 E. N. Bamberger, T. A. Harris, W. M. Kacmarsky, C. A. Moyer, R. J. Parker, J. J. Sherlock, and E. V. Zaretsky, *Life Adjustment Factors for Ball and Roller Bearings,* ASME, New York, 1971.

chapter **4** *JOURNAL BEARINGS*

THEO G. KEITH, Jr., Ph.D.
Professor and Chairman of Mechanical Engineering
University of Toledo
Toledo, Ohio

LIST OF SYMBOLS

a	Axial-flow land width
a_f	Pad load coefficient
A	Area
b	Circumferential-flow land width
C	Clearance
C^*	Specific heat
D	Diameter
e	Eccentricity
f	Coefficient of friction
F_j	Friction on journal
h	Film thickness
h_0	Minimum film thickness
H	Dimensionless film thickness
J	Mechanical equivalent of heat
k	Permeability
L	Bearing width
M	Rotor mass at bearing
M_j	Frictional torque on journal
n	Number of pads or recesses
N	Revolutions per unit time
p	Pressure
p_a	Ambient pressure
p_0	Short-bearing pressure
p_r	Recess pressure
p_s	Supply pressure
p_∞	Long bearing pressure

$\bar{p}$	Dimensionless pressure
P	Unit loading
q	Volume flow rate per unit length
q_f	Flow factor
Q	Volume flow rate
Q_s	Side leakage flow rate
R	Radius of journal
R_b	Radius of bearing $= R + C$
s	Stiffness
S	Sommerfeld number $= (\mu N/P)(R/C)^2$
t	Time
t_p	Thickness of porous liner
T	Temperature
u,v,w	Velocity in x, y, z directions, respectively
U	Velocity of journal
W	Load
W_R	Load component directed along line of centers
W_T	Load component normal to line of centers
x,y,z	Rectangular coordinates
X	Dimensionless minimum-film-thickness parameter $= (h_0/R)[P/(2\pi N\mu)]^{1/2}$
Y	Dimensionless frictional torque parameter $= [M_j/(WR)][P/(2\pi N\mu)]^{1/2}$
α	Porous material slip coefficient
β	Included angle of partial bearing, porous bearing parameter
β_1	Angle from line of centers to leading edge of partial bearing
γ	Circumferential-flow parameter
ε	Eccentricity ratio e/c
θ	Angular position measured from line of centers
θ_1	Angular position to leading edge of film
θ_2	Angular position to zero pressure in film
θ_3	Angular position to trailing edge of film
θ_{cav}	Angular position to cavitation boundary
Λ	Bearing number $= (6\mu\omega/p_a)(R/C)^2$
λ	Ratio of heat conduction loss to heat generation rate, reduced bearing number $= \Lambda/6$
μ	Dynamic viscosity
ρ	Density
τ	Shear stress
ϕ	Attitude angle
ω	Angular velocity
ζ	Dimensionless axial dimension $= z/(L/2)$
Ω	Porous bearing parameter

4-1 INTRODUCTION

The design of journal bearings is of considerable importance to the development of rotating machinery. Journal bearings are essential machine components to compressors, pumps, turbines, internal-combustion engines, motors, generators, etc.

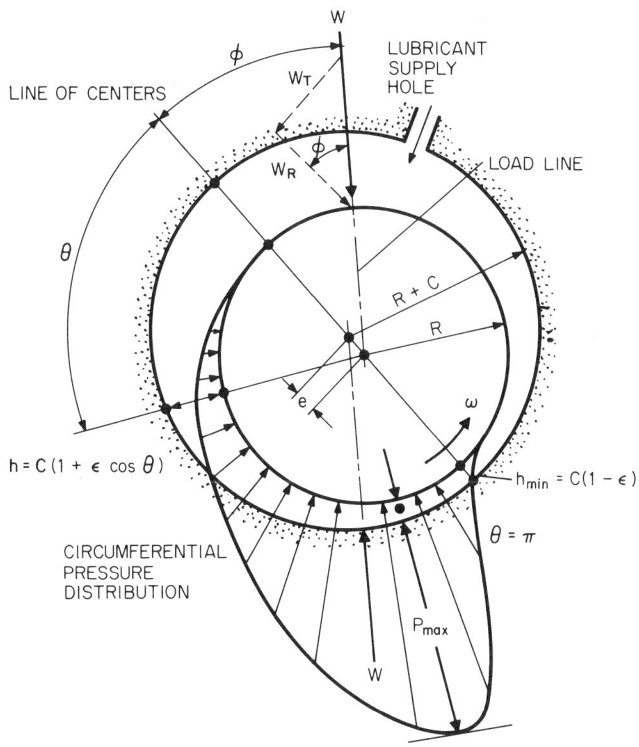

FIG. 4-1 Journal and bearing notation.

In its most basic form (Fig. 4-1), a *journal bearing* consists of a rotatable shaft (the journal) contained within a close-fitting cylindrical sleeve (the bearing). Generally, but not always, the bearing is fixed in a housing. The journal and bearing surfaces are separated by a film of lubricant (liquid or gas) that is supplied to the *clearance* space between the surfaces. The clearance space is generally quite small (on the order of one-thousandth of the journal radius) and has four major functions: to permit assembly of the journal and bearing, to provide space for the lubricant, to accommodate unavoidable thermal expansions, and to tolerate any shaft misalignment or deflection.

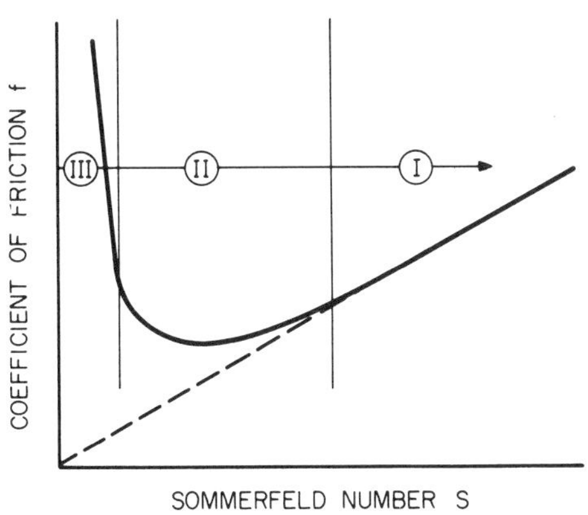

FIG. 4-2 Three lubrication regimes: I, thick film; II, thin film; III, boundary.

The fundamental purpose of a journal bearing is to provide radial support to a rotating shaft. Under load, the centers of the journal and the bearing are not coincident but are separated by a distance called the *eccentricity.* This eccentric arrangement establishes a converging-wedge geometry which, in conjunction with the relative motion of the journal and the bearing, permits a pressure to be developed by viscous effects within the thin film of lubricant and thus produces a load-carrying capability. However, if the load is too large or the shaft rotation too slow, the wedgelike geometry will not form and solid-to-solid contact can occur.

Journal bearings can operate in any of three lubrication regimes: *thick-film lubrication, thin-film lubrication,* or *boundary lubrication.* Generally, thick-film operation is preferred. Figure 4-2 is a diagram of the three lubrication regimes. Table 4-1 provides some of the characteristics of each regime.

Journal bearings may be classified according to the fluid mechanism that establishes the film load capacity: *Hydrodynamic* journal bearings, also called *self-acting* bearings, depend entirely on the relative motion of the journal and the bearing to produce film pressure for load support. *Hydrostatic* journal bearings, also called *externally pressurized* bearings, achieve load support by the supply of fluid from an external high-pressure source and require no relative motion between journal and bearing surfaces. *Hybrid* journal bearings are designed to use both hydrodynamic and hydrostatic principles to achieve load support between moving surfaces.

TABLE 4-1 Characteristics of Lubrication Regimes

Lubrication regime	Contact of bearing surfaces	Range of film thickness, in	Coefficient of friction	Degree of wear	Comments
Thick film	Only during startup or stopping	10^{-3}–10^{-4}	0.01–0.005	None	1. Light-loading high-speed regime 2. Friction coefficient proportional to $\mu N/[W/(LD)]$
Thin film	Intermittent; dependent on surface roughness	10^{-4} to 0.5×10^{-4}	0.005–0.05	Mild	1. High operating temperatures
Boundary	Surface to surface	0.5×10^{-4} to molecular thicknesses	0.05– 0.15	Large	1. Heavy-loading (unit load > 3000 psi) low-speed (< 60 fpm) operating regime 2. Heat generation and friction not dependent on lubricant viscosity

Notes ▪ Drawings ▪ Ideas

4-2 BEARING AND JOURNAL CONFIGURATIONS

4-2-1 Bearing Geometries

A wide range of bearing configurations are available to the journal bearing designer. Figure 4-3 depicts several of these bearings. The configurations range from the very simple plain journal bearing to the very complex tilting-pad bearing. The choice of bearing configuration depends on several factors. Among the more important are cost, load, power loss, dynamic properties, ease of construction, and difficulty of installation.

Journal bearings are termed *full bearings* (Fig. 4-3*a*) when the bearing surface completely surrounds the journal. Because they are easy to make and do not cost much, full bearings are the most commonly used bearing in rotating machinery. Full bearings become distorted during installation, and so they are generally not perfectly circular.

Journal bearings are called *partial bearings* when the bearing surface extends over only a segment of the circumference, generally 180° or less (Fig. 4-3*b*). Partial bearings are used in situations where the load is mainly unidirectional. Partial journal bearings have been found to reduce frictional torque on the journal and provide convenient accessibility, and they do not, in many instances, require strict manufacturing tolerance. Partial journal bearings in which the bearing radius exceeds the journal radius are called *clearance bearings,* while partial journal bearings in which the bearing and the journal radii are equal are termed *fitted bearings*.

Geometries in which two circular sectors are employed are called *elliptical,* or *lemon, bearings* (Fig. 4-3*c*). These bearings are really not elliptical at all but are fabricated by uniting two halves of a circular bearing which have had their mating

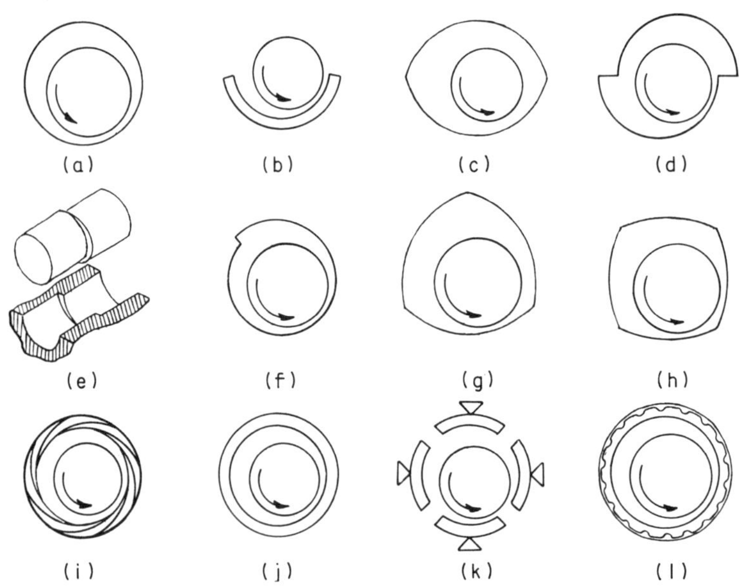

FIG. 4-3 Journal bearing geometries. (*a*) Full bearing; (*b*) partial bearing; (*c*) elliptical, or lemon, bearing; (*d*) offset bearing; (*e*) rocking journal bearing; (*f*) pressure dam bearing; (*g*) three-lobe bearing; (*h*) four-lobe bearing; (*i*) multileaf bearing; (*j*) floating-ring bearing; (*k*) tilting- or pivoted-pad bearing; (*l*) foil bearing.

faces machined so that the bearing has an approximately elliptical appearance. Lemon bearings are probably the most widely used bearing at low and moderate speeds. They are extensively used in turbine applications.

Elliptical bearings in which the two cylindrical halves are laterally displaced along the major axis are termed *offset bearings* (Fig. 4-3*d*). The relative displacement of the centers of each half of the bearing is called the *preset*. When the upper half of the bearing is displaced horizontally in the direction of rotation, the bearing has negative preset. It is found that load capacity increases with preset. Offset bearings have relatively high horizontal stiffness which helps prevent dynamic instability. Further, offset bearings allow greater lubricant flow and so run cooler.

Novel offset journal bearing designs for reducing power loss and wear in duty cycles which combine nonreversing loading with limited journal angular oscillation or in steady operation with counterrotation of journal and bearing under a constant load have been studied. In these applications conventional journal bearings are found to develop extremely thin lubricant films, which in turn results in high friction and wear. Figure 4-3*e* depicts a journal bearing in which both the journal and the bearing are divided axially into segments with offset centerlines. This arrangement produces a dynamic rocking motion which promotes a thicker lubricating film. Accordingly the assembly has been called a *rocking journal bearing*.

When a step is milled from the surface of the bearing (Fig. 4-3*f*), the resulting bearing is called a *pressure dam*, or *step, bearing*. The purpose of the step is to create additional hydrodynamic pressure on the top of the journal as the lubricant is rotated into the step. In turn, this pressure buildup enhances the load on the journal and therefore diminishes its susceptibility to vibration problems. Pressure dam bearings are very popular in the petrochemical industry.

Bearing geometries consisting of three or more sectors (Fig. 4-3*g* and *h*) are termed *lobed*, or *multilobed, bearings*. Generally, bearings with more than three lobes are used only in gas bearing applications. Multilobe bearings act as a number of partial bearings in series. The cost of multilobed bearings is considered moderate.

The *multileaf journal bearing* (Fig. 4-3*i*) is a variant of a multilobe bearing. It consists of a number of identical circular arcs, or leaves, whose centers are equally spaced around the generating circle. The operating characteristics of a multileaf bearing are practically independent of the direction of loading for bearings with eight or more leaves.

In a *floating-ring journal bearing* (Fig. 4-3*j*) the lubricating film is divided in two by the addition of a "floating" ring between the journal and the bearing. Floating-ring bearings have lower frictional losses and reduced heat generation and provide better stability.

Hydrodynamic journal bearings may be distinguished as to whether the bearing surface can pivot. The basic advantage of *pivoting*, or *tilting-pad, journal bearings* (Fig. 4-3*k*) over fixed-pad journal bearings is that they can accommodate, with little loss in performance, any shaft deflection or misalignment.

A *foil journal bearing* (Fig. 4-3*l*) consists of a very thin compliant bearing surface resting atop a series of corrugations. When it is compared to a conventional gas bearing, the foil bearing has a thicker film, higher load capacity, lower power loss, better stability, and superior endurance to high operating temperatures.

4-2-2 Journal Shapes

Although the journal is generally assumed to be perfectly circular, wear effects or poor manufacture can lead to journals with shapes as shown in Fig. 4-4*a*, *b*, and *c*. In addition, the possibility of developing pressure by grooving the surface of the jour-

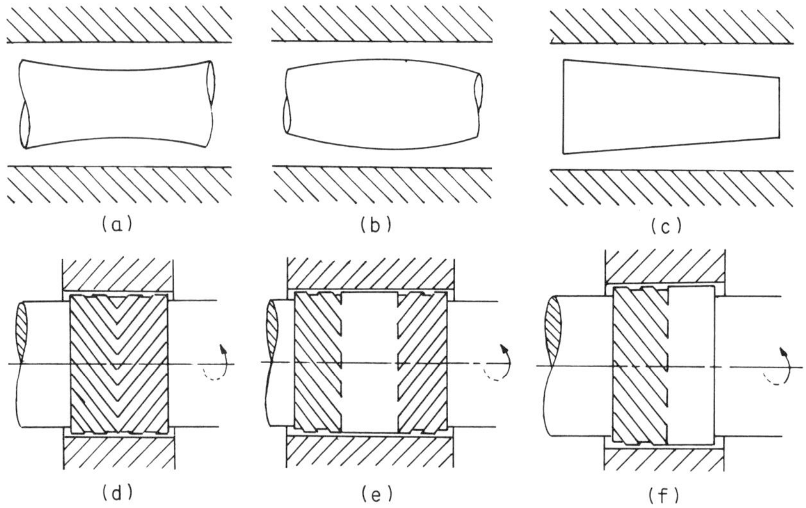

FIG. 4-4 Journal shapes. (*a*) Hourglass; (*b*) barrel; (*c*) tapered; (*d*) herringbone; (*e*) partly grooved symmetrical pattern; (*f*) partly grooved asymmetrical pattern. *(Parts (d), (e), and (f) are from [4-1].)*

nal has been investigated. Three grooved patterns that were found to yield good stability characteristics are shown in Fig. 4-4*c*, *d*, and *e*.

4-3 BEARING MATERIALS AND SELECTION CRITERIA

4-3-1 Bearing Materials

The ideal journal bearing material would have the following characteristics:

1. High compressive strength to withstand the applied radial loading
2. High fatigue strength to endure any cyclic changes in load direction and/or load intensity
3. Compatibility with the journal material to minimize surface scoring and bearing seizure whenever the journal and bearing surfaces come into contact (e.g., during startup)
4. Embedability to permit foreign particles in the lubricant to penetrate the bearing surface to avoid scoring and wear
5. Conformability of surface to tolerate journal misalignment, deflection, or manufacturing inaccuracies
6. High corrosion resistance to withstand chemical attack by the lubricant
7. High thermal conductivity to permit generated heat to be transported from the lubricant film
8. Appropriate coefficient of thermal expansion to avoid differences in thermal expansion of the journal and bearing
9. Low wear to prevent surface destruction especially under boundary lubrication

Notes ▪ Drawings ▪ Ideas

conditions (i.e., thin-film high-friction lubrication) and thereby lengthen the life of the bearing

Besides all these, the material should be inexpensive, highly available, and easily machined.

To be sure, no single material has been developed that satisfactorily combines all characteristics of the ideal bearing material. In fact, some of the characteristics are contradictory. For example, soft bearing materials generally do not have sufficient strength. To strengthen soft bearing materials, they are frequently bonded to stronger backing materials. Bearing linings or overlays may be cast, electrodeposited, sprayed, or chemically applied, and they have thicknesses which range from 0.01 to 0.5 inch (in).

Journal bearing materials may be broadly divided into two groups: metallics and nonmetallics. The metallic group includes aluminum alloys, babbitts (tin-, lead-, and aluminum-based), copper alloys (brass and bronze), zinc, and iron. The nonmetallic group includes plastics, carbon graphites, cemented carbides, and other proprietary materials. The nonmetallics have been widely used in self-lubrication applications because they can provide low friction and wear without the aid of a lubricant.

Because of the wide diversity of materials available for use in journal bearings, it is difficult to provide comprehensive tables of all relevant properties. Manufacturers and materials suppliers are the best sources for that information. Nevertheless, some physical properties of a variety of journal bearing materials are presented in Table 4-2 [4-2]. Typical applications and useful comments concerning a number of journal bearing alloys are displayed in Table 4-3, while Table 4-4 contains a numerical ranking of the performance characteristics of these alloys.

General information for a variety of self-lubricating materials is given in Table 4-5 [4-3]. Note that the table contains maximum values of the *PV* factor. This factor is the product of the bearing load per unit of projected area and the sliding velocity (i.e., speed in revolutions per minute times the bearing circumference). The *PV* parameter provides an indication of material wear and internal heat generation. Failure in self-lubricated bearings is frequently the direct result of internal overheating.

4-3-2 Bearing Material Selection Criteria

Selection of a bearing material invariably requires a compromise based on particular characteristics regarded by the designer to be of principal importance to the application at hand. DeGee [4-4] has developed a systematic approach for selecting a material for lubricated journal bearings. In this method, certain component criteria are identified within major property groups. Table 4-6 gives one such listing.

Not all the criteria presented in Table 4-6 need be considered. For example, in a particular application environmental properties may be of no concern because the bearing operates in a clean, moderate temperature environment and is not part of an electric machine.

After the list of criteria has been established, each component criterion is compared with all other criteria, and a graduation mark is allocated from 0, if there is no difference in the criteria, to 3, if there are large differences. For example, compressive strength (A1 in Table 4-6) might receive a 0 when compared with fatigue strength (A2) but receive a 3 when compared to thermal conductivity (B1), and so forth. When all component criteria have been compared with one another and graduation marks assigned, the graduation marks of each criterion are totaled and the sum of

Notes ▪ Drawings ▪ Ideas

TABLE 4-2 Physical Properties of Journal Bearing Materials

Material	Hardness H_B	Tensile strength, kpsi	Modulus of elasticity, Mpsi	Thermal conductivity, Btu/(h·ft·°F)	Coefficient of expansion, μin/(in·°F)	Density, lbm/ft³
Metals						
Lead babbitt	21	10	4.2	14	14	630
Tin babbitt	25	11	7.6	32	13	462
Copper lead	25	8	7.6	170	11	562
Silver	25	23	11	238	10.9	655
Cadmium	35	....	8	53	16.6	537
Aluminum alloy	45	22	10.3	119	13.5	181
Lead bronze	60	34	14	27	9.9	555
Tin bronze	70	45	16	29	10	549
Steel	150	75	30	29	6.4	487
Cast iron	180	35	23	30	5.7	449
Porous metals						
Bronze	40	18	....	17	10.5	399
Iron	50	25	....	16	6.7	381
Aluminum	H55†	15	....	....	....	144
Plastics						
TFE	D60‡	3	0.06	0.10	55	137
Nylon	M79†	11	0.41	0.14	55	71
Phenolic	M100	10	0.5	0.21	12	85
Acetal	M94	10	0.41	0.13	45	89
Polycarbonate	M70	8.5	0.32	0.11	70	75
Filled polyimide	E99†	7.5	....	0.44	22	89
Other nonmetallics						
Rubber	....	....	....	0.09	43	75
Wood	....	1.1	1.8	0.11	2.7	42
Carbon graphite	75§	2	2	10	1.5	106
Cemented tungsten carbide	A91†	130	81	40	3.3	886
Fused aluminum oxide	A85	30	50	1.6	8.2	243

†Rockwell. ‡Shore durometer. §Shore scleroscope.

SOURCE: Ref. [4-2].

Notes ▪ Drawings ▪ Ideas

TABLE 4-3 Bearing Alloy Material Applications

Material	Nominal composition, % by weight	Applications and remarks
Aluminum, low tin	Al 92 Sn 8	Tin added to improve compatibility; too much tin lowers strength. Has thermal expansion problems in steel housings. Requires hard journals. Good at high temperatures. Used in diesel engines and compressors.
Aluminum, high tin	Al 80 Sn 20	Produced by special working and annealing process so tin content does not greatly reduce strength. Used in automotive engines (crankshafts) and in aircraft equipment.
Babbitt, tin-based	Sn 84 Cu 8 Sb 8	Fatigue strength decreases as thickness increases. Low load capacity, thus usually bonded to one (bimetal) or two (trimetal) backing materials. Good in dirty applications, motors.
Babbitt, lead-based	Pb 75 Sn 10 Sb 15	Antimony (Sb) greater than 15% can cause brittleness. Cheaper than tin-based babbitt. Used in crankshaft bearings, transmission bushings, and electric equipment.
Lead bronze	Cu 70 Pb 25 Sn 5	Good for high-load high-speed applications; can be used with soft journals. Used as bushings in pumps, many home appliances, railroad cars.
Phosphor bronze	Cu 80 Sn 10 Pb 10	General-duty popular bushing; tin added to improve strength. Has high hardness; should be used with harder journals (300 BHN). Good impact resistance; used in lathes, pumps, home appliances.
Copper lead (cast)	Cu 75 Pb 25	Lead in pockets in copper matrix. Lead improves bearing surface but has corrosion problems. Frequently used as lining material on steel-backed bearings. Used in heavy-duty applications.
Copper lead (sintered)	Cu 75 Pb 25	Frequently used with a babbitt overlay in a trimetal bearing. Widely used in heavy-duty (high-temperature high-load) applications.
Silver (oven-plated)		Frequently used with lead indium overlay.

all these totals is divided into each amount, to obtain the component criteria weighting factors. The sum of all the weighting factors obviously is unity.

Next the candidate materials are given quality marks for the various component criteria. These marks range from 5 (excellent, or high) to 1 (poor, or low). For instance, tin-based babbitts are known to have only fair (2) fatigue strength whereas they have excellent (5) resistance to corrosion. The final ranking of the candidate materials is obtained by comparing the sums of the products of all component criteria weighting factors and the quality marks.

Notes ▪ Drawings ▪ Ideas

TABLE 4-4 Performance Ratings from 5 (High) to 1 (Low) for Bearing Alloy Materials

Material	Nominal composition, % by weight	Fatigue strength	Corrosion resistance	Seizure resistance	Embedability	Compatibility	Thermal conductivity
Aluminum, low tin	Al 92 Sn 8	2	4	2	1	1	4
Aluminum, high tin	Al 80 Sn 20	2	4	2	2	2	4
Babbitt, tin-based	Sn 84 Cu 8 Sb 8	1	5	4	5	4	3
Babbitt, lead-based	Pb 75 Sn 10 Sb 15	1	3	4	5	4	2
Lead bronze	Cu 70 Pb 25 Sn 5	4	2	3	3	2	3
Phosphor bronze	Cu 80 Sn 10 Pb 10	4	2	3	1	2	3
Copper lead (cast)	Cu 75 Pb 25	3	1	3	3	3	4
Copper lead (sintered)	Cu 60 Pb 40	3	1	3	3	3	4
Silver (over-plated)		5	4	1	1	1	5

TABLE 4-5 General Information on Self-Lubricating Bearing Materials

Material	Maximum load,† kpsi	Maximum speed, fpm	Maximum *PV* factor,‡ kpsi·fpm	Friction coefficient	Critical temperature, °F	Resistance to humidity	Resistance to chemical§
Nylon	1.5	200–400	1	0.1–0.4	400	Fair	Good
Acetal	1.5	200–500	1	0.1–0.4	300	Good	Good
Polyimide	10	1000	0.3	0.1–0.3	600	Good	Good
Phenolic	4	1000	0.1	0.9–1.1	300–400	Good	Good
Filled nylon	2	200–400	1	0.1–0.4	400	Fair	Good
Acetal PTFE filled	1.8	800	2.5	0.05–0.15	300	Good	Good
Filled polyimide	10	1000	6	0.1–0.3	600	Good	Good
Reinforced phenolic	4–5	200	4	0.1–0.4	300–400	Good	Good
Filled PTFE	1	500–1000	5–20	0.05–0.25	500	Excellent	Excellent
PTFE	100	50	1–10	0.05–0.25		Excellent	Excellent

†Load on projected area at zero speed.
‡For continuous service.
§At bearing surface.
SOURCE: Ref. [4-3].

TABLE 4-6 Journal Bearing Material Selection Criteria

Major property group	Component criteria
A. Mechanical	1. Compressive strength 2. Fatigue strength 3. Conformability (modulus of elasticity)
B. Thermal	1. Thermal conductivity 2. Thermal expansion
C. Chemical	1. Corrosion rate
D. Manufacturing	1. Cost 2. Machinability 3. Availability of material
E. Environmental	1. Behavior under abrasive conditions (embedability) 2. Resistance against electric-discharge pitting 3. Resistance to thermal degradation
F. Tribological	1. Wear rate 2. Coefficient of friction 3. Cavitation erosion resistance

4-4 PRESSURE EQUATION FOR A LUBRICATING FILM

4-4-1 Reynolds Equation

The differential equation which governs the pressure in a lubricating film is called the *Reynolds equation.* Bearing performance can be evaluated once the solution of this equation is in hand.

To develop the Reynolds equation, consider a portion of the fluid film of a journal bearing (Fig. 4-1). In general, there are three velocity components in the film: u, v, and w. There are three equations of motion (momentum equations), one for each coordinate direction. The collection is known as the *Navier-Stokes equations,* and each equation may be written in the following form:

$$\text{Inertial forces} = \text{pressure forces} + \text{body forces} + \text{viscous forces} \tag{4-1}$$

The Navier-Stokes equations in their complete form are too involved for analytical solution. They can, however, be reduced, and subsequently solved, by making several simplifying yet plausible assumptions.

1. The flow is laminar.
2. The inertial and body forces are small compared to the pressure and viscous forces.
3. The curvature of the film is negligible; the bearing surfaces are, therefore, nearly parallel.
4. The variation of pressure across the film $\partial P/\partial y$ is negligibly small.
5. The transverse velocity component across the film, v, is small compared to the other velocity components.
6. The velocity gradients across the film dominate over all other velocity gradients.

Notes ▪ Drawings ▪ Ideas

Application of these assumptions to mathematical versions of Eq. (4-1), and to an integrated version of the conservation of mass (the continuity equation), yields the *Reynolds equation* for a liquid-lubricated bearing:

$$\frac{\partial}{\partial x}\left(\frac{h^3}{\mu}\frac{\partial p}{\partial x}\right) + \frac{\partial}{\partial z}\left(\frac{h^3}{\mu}\frac{\partial p}{\partial z}\right) = 6(U_b + U_j)\frac{\partial h}{\partial x} + 6h\frac{\partial(U_b + U_j)}{\partial x} + 12\frac{\partial h}{\partial t} \tag{4-2}$$

The first grouping on the right-hand side of Eq. (4-2) is called the *wedge* term and must be negative to generate positive pressures. The third term on the right is called the *squeeze* term, and it will generate positive pressures when $\partial h/\partial t < 0$. The squeeze term vanishes for a steadily loaded bearing. Both the wedge and squeeze terms vanish for a purely hydrostatic case. If the bearing surface is fixed ($U_b = 0$), if the shaft is rotating with a speed ω (that is, $U_j = U = R\omega$), and if the viscosity of the lubricant is constant, then Eq. (4-2) may be written as

$$\frac{\partial}{\partial x}\left(h^3\frac{\partial p}{\partial x}\right) + \frac{\partial}{\partial z}\left(h^3\frac{\partial p}{\partial z}\right) = 6\mu R\omega\frac{\partial h}{\partial x} + 6\mu h\frac{\partial U}{\partial x} + 12\mu\frac{\partial h}{\partial t} \tag{4-3}$$

For comparative purposes, it is useful to cast the steady version of Eq. (4-3) into nondimensional form. This can be accomplished by defining the following nondimensional variables:

$$\theta \equiv \frac{x}{R} \qquad \zeta \equiv \frac{z}{L/2} \qquad H \equiv \frac{h}{C}$$

$$\bar{p} \equiv \frac{p}{6\mu(U/R)}\left(\frac{C}{R}\right)^2 = \frac{p}{6\mu\omega}\left(\frac{C}{R}\right)^2$$

Also since $\omega = 2\pi N$,

$$\bar{p} = \frac{p}{12\pi\mu N}\left(\frac{C}{R}\right)^2$$

Substituting these into Eq. (4-3) yields

$$\frac{\partial}{\partial\theta}\left(H^3\frac{\partial\bar{p}}{\partial\theta}\right) + \left(\frac{D}{L}\right)^2\frac{\partial}{\partial\zeta}\left(H^3\frac{\partial\bar{p}}{\partial\zeta}\right) = \frac{\partial H}{\partial\theta} \tag{4-4}$$

This equation can be interpreted as

Circumferential pressure flow + axial pressure flow = shear flow

The governing equation of a gas film differs from that of a liquid film by the appearance of the *density* ρ. The steady compressible version of the Reynolds equation for an isoviscous gas can be written

$$\frac{\partial}{\partial x}\left(h^3\rho\frac{\partial p}{\partial x}\right) + \frac{\partial}{\partial z}\left(h^3\rho\frac{\partial p}{\partial z}\right) = 6\mu R\omega\frac{\partial(\rho h)}{\partial x} \tag{4-5}$$

Since the energy dissipated by frictional forces is very small in normal gas bearing operation, we may assume the film has a constant temperature and $p = k\rho$; thus, Eq. (4-5) becomes

$$\frac{\partial}{\partial x}\left(h^3\frac{\partial p^2}{\partial x}\right) + \frac{\partial}{\partial z}\left(h^3\frac{\partial p^2}{\partial z}\right) = 12\mu R\omega\frac{\partial(ph)}{\partial x} \tag{4-6}$$

Notes ▪ Drawings ▪ Ideas

This equation can be written in nondimensional form as

$$\frac{\partial}{\partial\theta}\left(H^3\frac{\partial\bar{p}^2}{\partial\theta}\right)+\left(\frac{D}{L}\right)^2\frac{\partial}{\partial\zeta}\left(H^3\frac{\partial\bar{p}^2}{\partial\zeta}\right)=2\Lambda\frac{\partial(\bar{p}H)}{\partial\theta} \tag{4-7}$$

where $\bar{p}\equiv\dfrac{p}{p_a}$ $\quad p_a$ = ambient or supply pressure

$$H\equiv\frac{h}{C}\qquad \zeta=\frac{z}{L/2}\qquad \Lambda\equiv\frac{6\mu\omega}{p_a}\left(\frac{R}{C}\right)^2$$

Here Λ is called the *bearing*, or *compressibility, number*.

To solve the Reynolds equation, we require an expression for the *film thickness* h. From the triangle ABC in Fig. 4-5 (page 144), we may write

$$\overline{AC}=h+R=R_b\cos\xi+e\cos\theta$$

where θ is measured from the line of centers. And since $e\ll R_b$, $\cos\xi\sim 1$, thus

$$h=R_b-R+e\cos\theta$$

The *radial clearance* is $C\equiv R_b-R$, and the *eccentricity ratio* is $\varepsilon\equiv e/C$, hence

$$h=C(1+\varepsilon\cos\theta) \tag{4-8}$$

4-4-2 Boundary Conditions

Three sets of circumferential boundary conditions are commonly applied to the solutions of the Reynolds equation. These boundary conditions have been given the

TABLE 4-7 Typical Boundary Conditions on the Reynolds Equation

Names associated with boundary conditions	Pressure profile	Mathematical expression
Sommerfeld (full Sommerfeld)	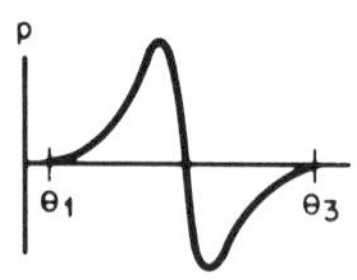	$p(\theta_1)=p(\theta_3)=0$ (zero pressure means ambient or atmospheric pressure) For complete journal bearings: $\theta_1=0,\ \theta_3=2\pi$ For partial journal bearings: $\theta_1=\beta_1,\ \theta_3=\theta_1+\beta$
Gumbel (half Sommerfeld)	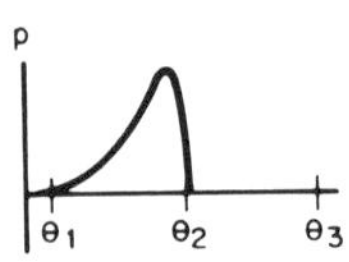	$p(\theta_1)=p(\theta_2)=0$ $p(\theta_2\le\theta\le\theta_3)=0$ For complete journal bearings: $\theta_1=0,\ \theta_2=\pi,\ \theta_3=2\pi$
Swift-Stieber (Reynolds)	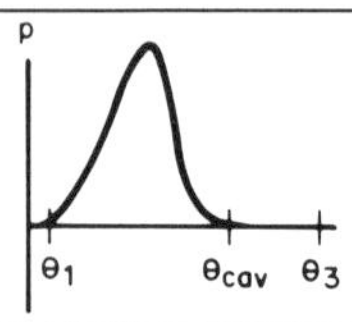	$p(\theta_2)=p_{cav}$ = atmospheric pressure $\dfrac{\partial p}{\partial\theta}(\theta_2)=0$ $\theta_2=\theta_{cav}$ which must be determined

Notes ▪ Drawings ▪ Ideas

names *Sommerfeld, Gumbel,* and *Swift-Stieber.* Of the three, the Sommerfeld conditions are the easiest to apply, but they yield certain unrealistic results. For example, in a liquid-lubricated bearing, Sommerfeld conditions produce negative pressures in the film. This results in the shaft being displaced at right angles to the load line as the load is increased. Gumbel conditions are similar to Sommerfeld conditions except that all negative pressures are disregarded. Although this approach leads to more realistic load results, it produces a violation of the conservation of mass. The Swift-Stieber conditions come closest to representing the actual conditions in a film, but they are more difficult to apply. They are widely used in numerical investigations. The three sets of boundary conditions are summarized in Table 4-7.

4-5 JOURNAL BEARING PERFORMANCE

Once the pressure distribution is established, the journal bearing performance may be determined. Performance is generally measured in terms of four quantities: bearing load capacity, frictional losses, lubricant flow requirement, and temperature rise.

4-5-1 Bearing Load Relations

Five parameters are associated with the load capacity of a journal bearing:

1. The radial load component W_R acts along the line of centers (Fig. 4-1) and is computed from

$$W_R = -R \int_{-L/2}^{L/2} \int_{\theta_1}^{\theta_3} p \cos \theta \, d\theta \, dz$$

where θ_1 and θ_3 are, respectively, the leading and trailing angular locations of the lubricating film.

2. The tangential load component W_T acts perpendicular to the line of centers:

$$W_T = R \int_{-L/2}^{L/2} \int_{\theta_1}^{\theta_3} p \sin \theta \, d\theta \, dz$$

3. The bearing load W must be supported by the pressure developed within the lubricating film. Generally the load is specified or enters the design via the *unit load P*, which is defined as the load per unit projected area, or

$$P = \frac{W}{LD}$$

Typical values of the unit load are given in Table 4-8.

4. The *Sommerfeld number S* is a dimensionless parameter that characterizes bearing performance; large S (say, greater than 0.15) indicates a lightly loaded bearing operating at a small eccentricity. The Sommerfeld number may be calculated from

$$S = \frac{\mu UL}{\pi W}\left(\frac{R}{C}\right)^2 = \frac{\mu N}{P}\left(\frac{R}{C}\right)^2$$

5. The *attitude angle* ϕ is the angular distance between the load line and the line of

Notes ▪ Drawings ▪ Ideas

TABLE 4-8 Range of Unit Loads for Various Applications

Application	Bearing	Unit load range, kpsi
Automotive engines	Main	0.6–0.8
	Crankpin	1.7–2.3
Diesel engines	Main	0.9–2.3
	Crankpin	1.1–2.3
	Wristpin	2.0–2.3
Steam turbines	Main	0.12–0.25
Air compressors	Main	0.14–0.28
	Crankpin	0.28–0.5
Centrifugal pumps	Shaft	0.1–0.18
Electric motors	Shaft	0.12–0.25

centers (Fig. 4-1). It locates the minimum film thickness as measured from the load line. Because $W_R = W \cos \phi$ and $W_T = W \sin \phi$,

$$\phi = \tan^{-1} \frac{W_T}{W_R}$$

4-5-2 Bearing Friction Relations

Four parameters are involved with the frictional behavior of a journal bearing:

1. The *shear stress* τ, acting on either the shaft or the bearing surface, consists of two terms: one is due to motion of the shaft (pure shear), and the other is due to the circumferential pressure distribution (pressure-induced shear):

$$\tau = \frac{\mu U}{h} \pm \frac{h}{2R} \frac{\partial p}{\partial \theta}$$

The plus sign corresponds to the shear stress on the journal surface; the minus sign, to the shear stress on the bearing surface.

2. The *frictional force* acting on the journal F_j is found by integrating the pure shear stress over the entire surface of the journal and the pressure-induced shear stress up to the trailing edge of the film. This yields

$$F_j = \left(\frac{2\pi}{1 - \varepsilon^2}\right)(\mu UL)\left(\frac{R}{C}\right) + \frac{\varepsilon W_T}{2}\left(\frac{C}{R}\right)$$

3. The *friction coefficient* f is the ratio of the journal frictional force to the bearing load

$$f = \frac{F_j}{W}$$

The *friction variable* is the product $(R/C)(f)$ and so may be written

$$\left(\frac{R}{C}\right)(f) = \left(\frac{2\pi^2}{1 - \varepsilon^2}\right)(S) + \frac{\varepsilon \sin \phi}{2}$$

Notes ▪ Drawings ▪ Ideas

4. The power that must be supplied to the journal to overcome friction is called the *frictional horsepower loss* HP, and it may be computed from

$$\mathrm{HP} = C_1 F_j U = C_2 f W R N$$

where C_1 and C_2 depend on the system of units. For F_j in pounds (lb) and U in inches per second (in/s), $C_1 = \frac{1}{6600}$ for W (lb), R (in), and N in revolutions per second (r/s), $C_2 = 2\pi C_1 = 9.51998 \times 10^{-4}$.

The relevant geometry for the execution of these tasks is depicted in Fig. 4-5.

4-5-3 Lubricant Flow Relations

Lubricant flow rates are needed to estimate the capacity of the lubricant supply system and to determine the cooling requirements of the bearing. This involves evaluation of the lubricant flow within the clearance space, the lubricant flow that leaks out the sides of the bearing, and the lubricant flow that is supplied to the bearing.

In general, volume flow rate per unit length is composed of a term due to surface motion (shear flow) and another due to the pressure (pressure-induced flow). Journal bearing circumferential and axial flows per unit length are

$$q_\theta = \frac{Uh}{2} - \frac{h^3}{12\mu R}\frac{\partial p}{\partial \theta} \qquad q_z = -\frac{h^3}{12\mu}\frac{\partial p}{\partial z} \tag{4-9}$$

The total flow rates may be found by integrating Eq. (4-9) across the bearing length for the circumferential flow rate and around the bearing circumference for the axial flow rate.

Assuming that lubricant is supplied in the unloaded portion of the bearing (Fig. 4-6), we see that the rate at which lubricant leaks out of the active portion of the film is

$$Q_{sa} = Q_1 - Q_2 \tag{4-10}$$

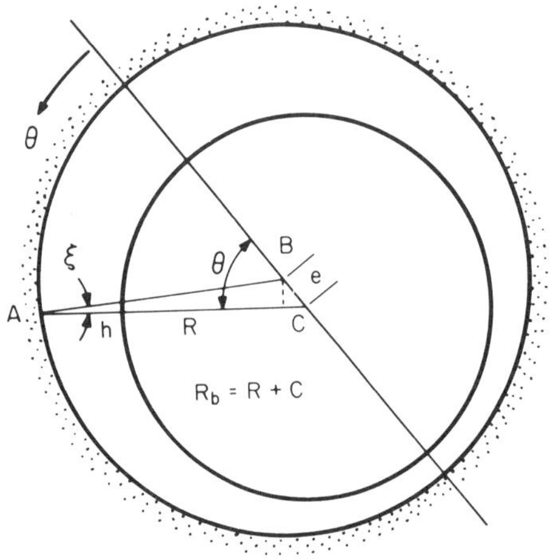

FIG. 4-5 Film thickness geometry

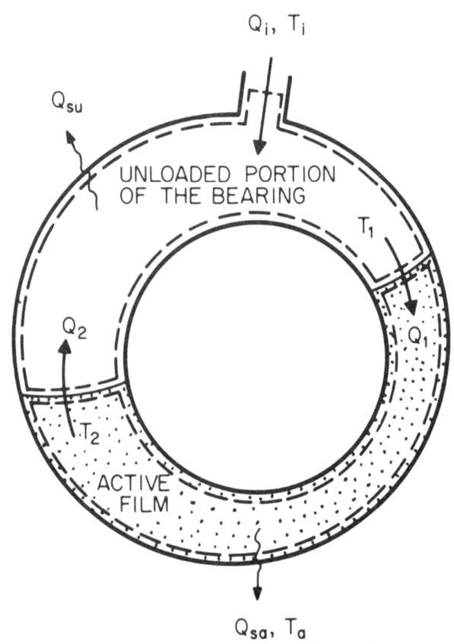

FIG. 4-6 Control volumes for lubricant flow and heat balances.

Notes ▪ Drawings ▪ Ideas

where Q_1 = flow into the leading edge of the film and Q_2 = flow out of the trailing edge.

When the input flow rate equals the leakage flow rate, Q_1 is called the *classical rate.* For given values of W, U, and μ, the classical rate is the largest flow that can be carried into the active film by shaft rotation. However, in practice Q_i, the input flow rate, may be greater (the *flooded condition*) or less (the *starved condition*) than the appropriate value to achieve the classical rate. For flooded conditions, side flow will also occur in the unloaded portion of the bearing, and we may write

$$Q_{su} = Q_2 + Q_i - Q_1 \tag{4-11}$$

thus,

$$Q_{su} + Q_{sa} = Q_s = Q_i$$

4-5-4 Bearing Thermal Relations

The steady energy balance equation can be simply expressed as

Energy inflow rate − energy outflow rate + energy generation rate = 0

An energy balance may be performed on the unloaded portion of the film (Fig. 4-6). Toward that end it is assumed that there is complete mixing between the inlet Q_i and the carryover Q_2 flows so that $T_u = T_1$. It is further assumed that there is no energy generation and negligible heat transfer. Hence, for the unloaded portion of the film,

$$Q_iT_i + Q_2T_2 = (Q_2 + Q_i)(T_1) \tag{4-12}$$

Next an energy balance is performed on the active portion of the lubricating film (Fig. 4-6). The energy generation rate is taken to be F_jU/J, and the conduction heat loss to the shaft and bearing are taken to be a portion of the heat generation rate, or $\lambda F_jU/J$. Accordingly,

$$\rho Q_1C^*T_1 - (\rho Q_{sa}C^*T_a + \rho Q_2C^*T_2) + \frac{(1-\lambda)F_jU}{J} = 0 \tag{4-13}$$

Combining Eqs. (4-10) to (4-13) and assuming that the side-flow leakage occurs at the average film temperature $T_a = (T_1 + 2\,T_2)/2$, we find that

$$\frac{J\rho C^*(T_a - T_i)}{(1-\lambda)P} = \frac{1 + 2Q_2/Q_i}{2 - Q_{sa}/Q_1}\,\frac{4\pi(R/C)(f)}{Q_1/(RCNL)} \tag{4-14}$$

This shows that the lubricant temperature rise is $1 - \lambda$ times the rise when conduction is neglected.

4-6 LIQUID-LUBRICATED JOURNAL BEARINGS

In the hydrodynamic operation of a liquid-lubricated journal bearing, it is generally assumed that the lubricant behaves as a continuous incompressible fluid. However, unless the lubricant is admitted to the bearing under relatively high hydrostatic head, the liquid film can experience periodic vaporization which can cause the film to rupture and form unstable pockets, or cavities, within the film. This disruption of the film is called *cavitation,* and it occurs when the pressure within the bearing falls to the vapor pressure of the lubricant. Narrow liquid-lubricated bearings are especially susceptible to this problem. Figure 4-7 illustrates the general film condition in which lubricant is admitted through a lubricating groove at some angular position θ_0. Clearly incomplete films complicate the analysis, and therefore the design, of a liquid-lubricated journal bearing.

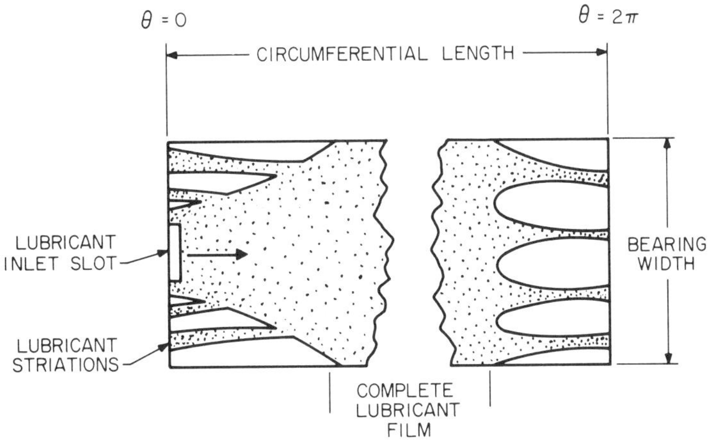

FIG. 4-7 Diagram of an incomplete fluid film.

4-6-1 *L/D* Effects on Cylindrical Full Journal Bearings

LONG-LENGTH BEARINGS. When the length of a bearing is such that $L > 2D$, the axial pressure flow term in the Reynolds equation may be neglected and the bearing performs as if it were infinitely long. Under this condition, the reduced Reynolds equation can be directly integrated. Table 4-9 contains long-bearing results for both Sommerfeld and Gumbel boundary conditions.

SHORT-LENGTH BEARINGS. When the length of a bearing is such that $L < D/4$, the axial pressure flow will dominate over the circumferential flow and again the Reynolds equation can be readily integrated. Results of such a short-bearing integration with Gumbel boundary conditions are shown in Table 4-10.

FINITE-LENGTH BEARINGS. The *slenderness ratio L/D* for most practical designs ranges between 0.5 and 2.0. Thus, neither the short-bearing theory nor the long-bearing theory is appropriate. Numerous attempts have been made to develop methods which simultaneously account for length and circumferential effects. Various analytical and numerical methods have been successfully employed. Although such techniques have produced important journal bearing design information, other simplified methods of analysis have been sought. These methods are useful because they do not require specialized analytical knowledge or the availability of large computing facilities. What is more, some of these simple, approximate methods yield results that have been found to be in good agreement with the more exact results. One method is described.

Reason and Narang [4-5] have developed an approximate technique that makes use of both long- and short-bearing theories. The method can be used to accurately design steadily loaded journal bearings on a hand-held calculator.

TABLE 4-9 Long-Bearing Pressure and Performance Parameters

Performance parameter	Sommerfeld conditions	Gumbel conditions
$\dfrac{p}{12\pi\mu N}\left(\dfrac{C}{R}\right)^2$	$\dfrac{(\varepsilon \sin\theta)(2+\varepsilon\cos\theta)}{(2+\varepsilon^2)(1+\varepsilon\cos\theta)^2}$	$\dfrac{(\varepsilon \sin\theta)(2+\varepsilon\cos\theta)}{(2+\varepsilon^2)(1+\varepsilon\cos\theta)^2}$ $0<\theta\ \ \pi,$ $0,\ \pi<\theta<2\pi$
$\dfrac{W_R}{3\mu UL}\left(\dfrac{C}{R}\right)^2$	0	$\dfrac{4\varepsilon^2}{(2+\varepsilon^2)(1-\varepsilon^2)}$
$\dfrac{W_T}{3\mu UL}\left(\dfrac{C}{R}\right)^2$	$\dfrac{4\pi\varepsilon}{(2+\varepsilon^2)\sqrt{1-\varepsilon^2}}$	$\dfrac{2\pi\varepsilon}{(2+\varepsilon^2)\sqrt{1-\varepsilon^2}}$
ϕ	$\dfrac{\pi}{2}$	$\tan^{-1}\left(\dfrac{\pi}{2}\dfrac{\sqrt{1-\varepsilon^2}}{\varepsilon}\right)$
S	$\dfrac{(2+\varepsilon^2)\sqrt{1-\varepsilon^2}}{12\pi^2\varepsilon}$	$\dfrac{(2+\varepsilon^2)(1-\varepsilon^2)}{6\pi\varepsilon\sqrt{4\varepsilon^2+\pi^2(1-\varepsilon^2)}}$
$\dfrac{F_j}{\mu UL}\dfrac{C}{R}$	$\dfrac{4\pi(1+2\varepsilon^2)}{(2+\varepsilon^2)\sqrt{1-\varepsilon^2}}$	$\dfrac{\pi(4+5\varepsilon^2)}{(2+\varepsilon^2)\sqrt{1-\varepsilon^2}}$
$\left(\dfrac{R}{C}\right)(f)$	$\dfrac{1+2\varepsilon^2}{3\varepsilon}$	$\dfrac{4+5\varepsilon^2}{6\varepsilon}\sqrt{\dfrac{1-\varepsilon^2}{4\varepsilon^2+\pi(1-\varepsilon^2)}}$
$\dfrac{Q_s}{RCNL}$	0	0

It was proposed that the film pressure p be written as a harmonic average of the short-bearing pressure p_0 and the long-bearing pressure p_∞, or

$$\frac{1}{p}=\frac{1}{p_0}+\frac{1}{p_\infty} \qquad \text{or} \qquad p=\frac{p_0}{1+p_0/p_\infty}$$

The pressure and various performance parameters that can be obtained by this combined solution approximation are presented in Table 4-11. Note that several of these parameters are written in terms of two quantities, I_s and I_c. Accurate values of these quantities and the Sommerfeld number are displayed in Table 4-12. With the exception of the entrainment flow, which is increasingly overestimated at large ε and L/D, the predictions of this simple method have been found to be very good.

EXAMPLE 1. Using the Reason and Narang combined solution approximation, determine the performance of a steadily loaded full journal bearing for the following conditions:

Notes ▪ Drawings ▪ Ideas

TABLE 4-10 Short-Bearing Pressure and Performance Parameters

Performance parameter	Gumbel conditions
$\dfrac{p}{12\pi\mu N}\left(\dfrac{C}{R}\right)^2$	$-\dfrac{\varepsilon\sin\theta}{2(1+\varepsilon\cos\theta)^3}\left(\dfrac{L}{D}\right)^2(\zeta^2-1)$ $\quad 0\le\theta\le\pi,$ $\quad 0,\ \pi\le\theta\le 2\pi$
$\dfrac{W_R}{3\mu UL}\left(\dfrac{C}{R}\right)^2$	$\dfrac{4\varepsilon^2}{3(1-\varepsilon^2)^2}\left(\dfrac{L}{D}\right)^2$
$\dfrac{W_T}{3\mu UL}\left(\dfrac{C}{R}\right)^2$	$\dfrac{\pi\varepsilon^2}{3(1-\varepsilon^2)^3}\left(\dfrac{L}{D}\right)^2$
$\dfrac{W}{3\mu UL}\left(\dfrac{C}{R}\right)^2$	$\dfrac{\varepsilon\sqrt{\pi^2(1-\varepsilon^2)+16\varepsilon^2}}{3(1-\varepsilon^2)^2}\left(\dfrac{L}{D}\right)^2$
ϕ	$\tan^{-1}\dfrac{\pi(1-\varepsilon^2)^2}{4\varepsilon}$
S	$\dfrac{(1-\varepsilon^2)^2}{\pi\varepsilon\sqrt{\pi^2(1-\varepsilon^2)+16\varepsilon^2}}\left(\dfrac{D}{L}\right)^2$
$\dfrac{F_j}{\mu UL}\dfrac{C}{R}$	$\dfrac{2\pi}{\sqrt{1-\varepsilon^2}}$
$\left(\dfrac{R}{C}\right)(f)$	$\dfrac{(2\pi)(1-\varepsilon^2)^{3/2}}{\varepsilon\sqrt{\pi^2(1-\varepsilon^2)+16\varepsilon^2}}$
$\dfrac{Q_s}{RCNL}$	$2\pi\varepsilon$

$\mu = 4\times10^{-6}$ reyn $\qquad D = 1.5$ in
$N = 1800$ rpm $\qquad L = 1.5$ in
$W = 500$ lbf $\qquad C = 1.5\times10^{-3}$

Solution. The unit load is $P = W/(LD) = 222$ pounds per square inch (psi), and the Sommerfeld number is

$$S = \frac{\mu N}{P}\left(\frac{R}{C}\right)^2 = 0.135$$

Entering Table 4-12 at this Sommerfeld number and a slenderness ratio of 1, we find that $\varepsilon = 0.582$, $I_c = 0.2391$, and $I_s = 0.3119$. The bearing performance is computed by evaluating various parameters in Table 4-11. Results are compared in Table 4-13 to values obtained by Shigley and Mitchell [4-6] by using design charts.

Notes ▪ Drawings ▪ Ideas

TABLE 4-11 Pressure and Performance Parameters of the Combined Solution Approximation

Performance parameter	Equation
$\dfrac{p}{12\pi\mu N}\left(\dfrac{C}{R}\right)^2$	$\dfrac{\dfrac{1}{2}\left(\dfrac{L}{D}\right)^2(1-\zeta^2)\dfrac{\varepsilon\sin\theta}{(1+\varepsilon\cos\theta)^3}}{1+\left(\dfrac{L}{D}\right)^2\dfrac{(2+\varepsilon^2)(1-\zeta^2)}{2(1+\varepsilon\cos\theta)(2+\varepsilon\cos\theta)}}$
$\dfrac{W_R}{3\mu UL}\left(\dfrac{C}{R}\right)^2$	$-2I_c$
$\dfrac{W_T}{3\mu UL}\left(\dfrac{C}{R}\right)^2$	$2I_s$
ϕ	$\tan^{-1}\left(\dfrac{-I_s}{I_c}\right)$
S	$\dfrac{1}{6\pi\sqrt{I_s^2+I_c^2}}$
$\dfrac{F_j}{\mu UL}\dfrac{C}{R}$	$3\varepsilon I_s+\dfrac{2\pi}{\sqrt{1-\varepsilon^2}}$
$\left(\dfrac{R}{C}\right)(f)$	$6\pi S\left(\dfrac{\varepsilon I_S}{2}+\dfrac{\pi}{3\sqrt{1-\varepsilon^2}}\right)$
$\dfrac{Q_{0,\pi}}{RCNL}$ †	$\pi\left[1+\varepsilon\mp\varepsilon E\left(1-\dfrac{2E}{\sqrt{1+2E}}\tanh^{-1}\dfrac{1}{\sqrt{1+2E}}\right)\left(\dfrac{L}{D}\right)^2\right]$ where $E=\dfrac{(1\pm\varepsilon)(2\pm\varepsilon)}{(2+\varepsilon^2)}\left(\dfrac{D}{L}\right)^2$
$\dfrac{Q_s}{Q_0}$	$1-\dfrac{Q_\pi}{Q_0}$
$\dfrac{J\pi C^*\,\Delta T}{P}$	$\dfrac{1}{1-\frac{1}{2}Q_s/Q_0}\,\dfrac{4\pi(R/C)f}{Q_0/(RCNL)}$

†For Q_0 (flow through maximum film thickness at $\theta = 0$) use top signs; for Q_π (flow through minimum film thickness at $\theta = \pi$) use lower signs.

Notes ▪ Drawings ▪ Ideas

TABLE 4-12 Values of I_s, I_c, and Sommerfeld Number for Various Values of L/D and ε

ε \ L/D	0.25	0.5	0.75	1.0	1.5	2	∞
0.1	0.0032† −0.0004 16.4506	0.0120 −0.0014 4.3912	0.0244 −0.0028 2.1601	0:0380 −0.0041 1.3880	0.0636 −0.0063 0.8301	0.0839 −0.0076 0.6297	0.1570 −0.0100 0.3372
0.2	0.0067 −0.0017 7.6750	0.0251 −0.0062 2.0519	0.0505 −0.0118 1.0230	0.0783 −0.0174 0.6614	0.1300 −0.0259 0.4002	0.1705 −0.0312 0.3061	0.3143 −0.0408 0.1674
0.3	0.0109 −0.0043 4.5276	0.0404 −0.0153 1.2280	0.0804 −0.0289 0.6209	0.1236 −0.0419 0.4065	0.2023 −0.0615 0.2509	0.2628 −0.0733 0.1944	0.4727 −0.0946 0.1100
0.4	0.0164 −0.0089 2.8432	0.0597 −0.0312 0.7876	0.1172 −0.0579 0.4058	0.1776 −0.0825 0.2709	0.2847 −0.1183 0.1721	0.3649 −0.1391 0.1359	0.6347 −0.1763 0.0805
0.5	0.0241 −0.0174 1.7848	0.0862 −0.0591 0.5076	0.1656 −0.1065 0.2694	0.2462 −0.1484 0.1845	0.3835 −0.2065 0.1218	0.4831 −0.2391 0.0984	0.8061 −0.2962 0.0618
0.6	0.0363 −0.0338 1.0696	0.1259 −0.1105 0.3167	0.2345 −0.1917 0.1752	0.3306 −0.2590 0.1242	0.5102 −0.3474 0.0859	0.6291 −0.3949 0.0714	0.9983 −0.4766 0.0480
0.7	0.0582 −0.0703 0.5813	0.1927 −0.2161 0.1832	0.3430 −0.3549 0.1075	0.4793 −0.4612 0.0798	0.6878 −0.5916 0.0585	0.8266 −0.6586 0.0502	1.2366 −0.7717 0.0364
0.8	0.1071 −0.1732 0.2605	0.3264 −0.4797 0.0914	0.5425 −0.7283 0.0584	0.7220 −0.8987 0.0460	0.9771 −0.0941 0.0362	1.1380 −1.1891 0.0322	1.5866 −0.3467 0.0255
0.9	0.2761 −0.6644 0.0737	0.7079 −1.4990 0.0320	1.0499 −2.0172 0.0233	1.3002 −2.3269 0.0199	1.6235 −2.6461 0.0171	1.8137 −2.7932 0.0159	2.3083 −3.0339 0.0139
0.95	0.6429 −2.1625 0.0235	1.3712 −3.9787 0.0126	1.8467 −4.8773 0.0102	2.1632 −5.3621 0.0092	2.5455 −5.8315 0.0083	2.7600 −6.0396 0.0080	3.2913 −6.3776 0.0074
0.99	3.3140 −22.0703 0.0024	4.9224 −28.5960 0.0018	5.6905 −30.8608 0.0017	6.1373 −31.9219 0.0016	6.6295 −32.8642 0.0016	6.8881 −33.2602 0.0016	8.7210 −33.5520 0.0015

†The three numbers associated with each ε and L/D pair are, in order from top to bottom, I_s, I_c, and S.

TABLE 4-13 Comparison of Predicted Performance between Two Methods for Example 1

Method \ Parameter	ε	ϕ	$\left(\frac{R}{C}\right)(f)$	$\frac{Q}{RCNL}$	$\frac{Q_s}{Q}$	ΔT
Combined solution approximation	0.582	52.5°	3.508	4.473	0.652	26.6°F
Design charts†	0.58	53.°	3.50	4.28	0.655	26.6°F

†SOURCE: Shigley and Mitchell [4-6].

4-6-2 Design Charts

Design charts have been widely used for convenient presentation of bearing performance data. Separate design graphs are required for every bearing configuration or variation. Use of the charts invariably requires repeated interpolations and extrapolations. Thus, design of journal bearings from these charts is somewhat tedious.

RAIMONDI-BOYD CHARTS. The most famous set of design charts was constructed by Raimondi and Boyd [4-7]. They presented 45 charts and 6 tables of numerical information for the design of bearings with slenderness ratios of 1/4, 1/2, and 1 for both partial (60°, 120°, and 180°) and full journal bearings. Consequently, space does not permit all those charts to be presented. Instead a sampling of the charts for bearings with an L/D ratio of 1 is given. Figures 4-8 to 4-13 present graphs of the *minimum-film-thickness variable* h_0/C (note that $h_0/C = 1 - \varepsilon$),the *attitude angle* ϕ (or location of the minimum thickness), the *friction variable* $(R/C)(f)$, the *flow variable* $Q/(RCNL)$, the *flow ratio* Q_s/Q, and the *temperature-rise variable* $J\gamma C^* \Delta T/P$. Table 4-14 is a tabular presentation of these data.

For slenderness ratios other than the four displayed (∞, 1, 1/2, and 1/4), Raimondi and Boyd suggest the use of the following interpolation formula:

$$y_{L/D} = \left(\frac{D}{L}\right)^3 \left[-\frac{1}{8}\left(1 - \frac{L}{D}\right)\left(1 - 2\frac{L}{D}\right)\left(1 - 4\frac{L}{D}\right) y_\infty + \frac{1}{3}\left(1 - 2\frac{L}{D}\right)\left(1 - 4\frac{L}{D}\right) y_1 - \frac{1}{4}\left(1 - \frac{L}{D}\right)\left(1 - 4\frac{L}{D}\right) y_{1/2} + \frac{1}{24}\left(1 - \frac{L}{D}\right)\left(1 - 2\frac{L}{D}\right) y_{1/4} \right]$$

where y = any performance variable, that is, $(R/C)(f)$, h_0/C, etc., and the subscript of y is the L/D value at which the variable is being evaluated.

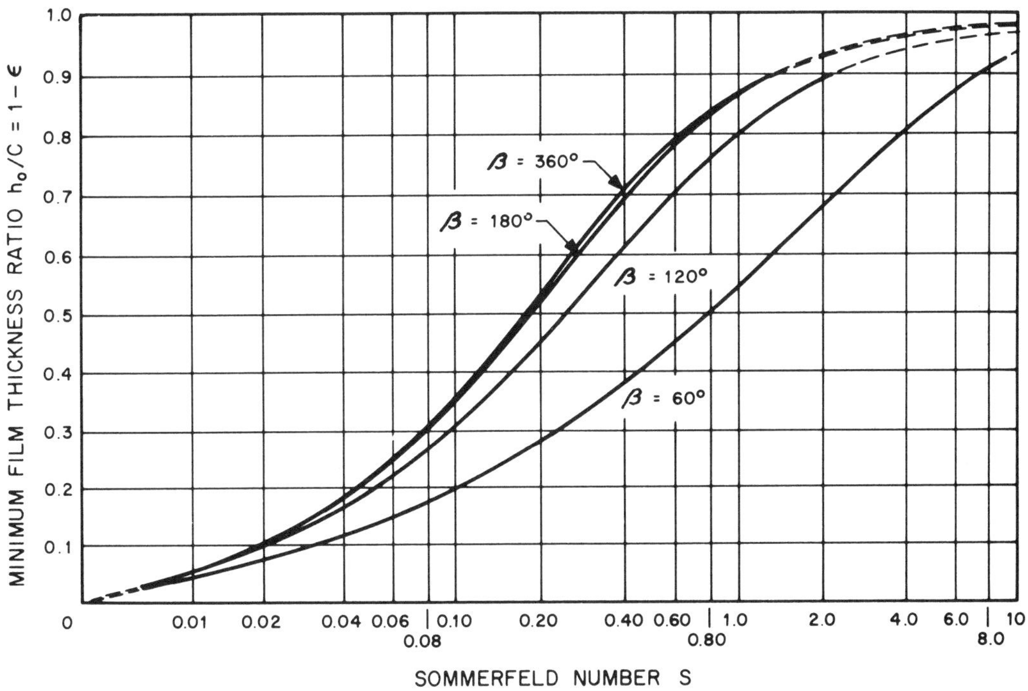

FIG. 4-8 Minimum film thickness ratio versus Sommerfeld number for full and partial journal bearings, $L/D = 1$, Swift-Stieber boundary conditions. *(From Raimondi and Boyd [4-7].)*

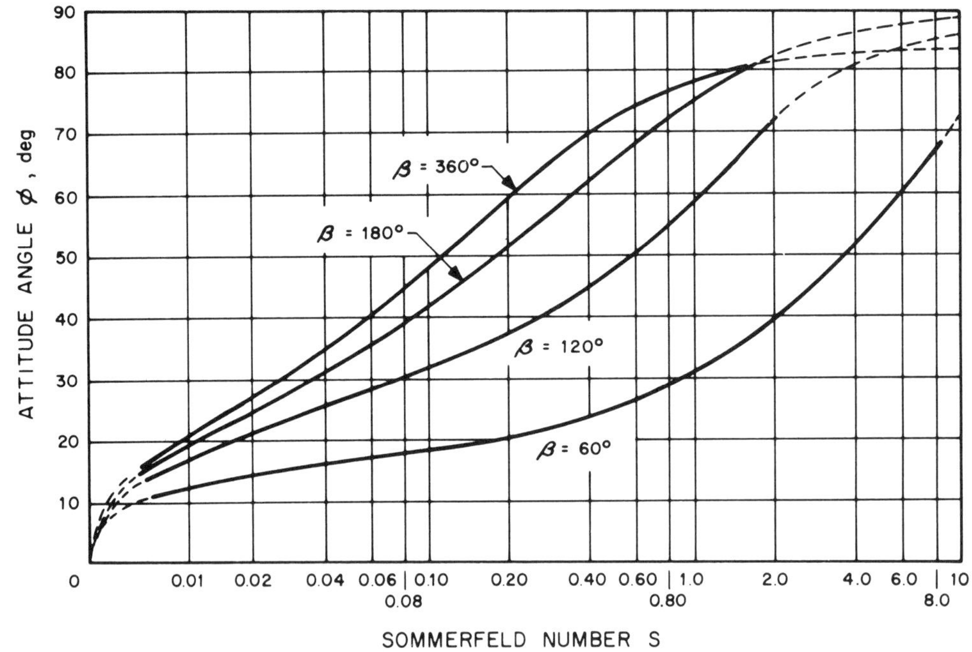

FIG. 4-9 Attitude angle versus Sommerfeld number for full and partial journal bearings, $L/D = 1$, Swift-Stieber boundary conditions. *(From Raimondi and Boyd* [4-7].*)*

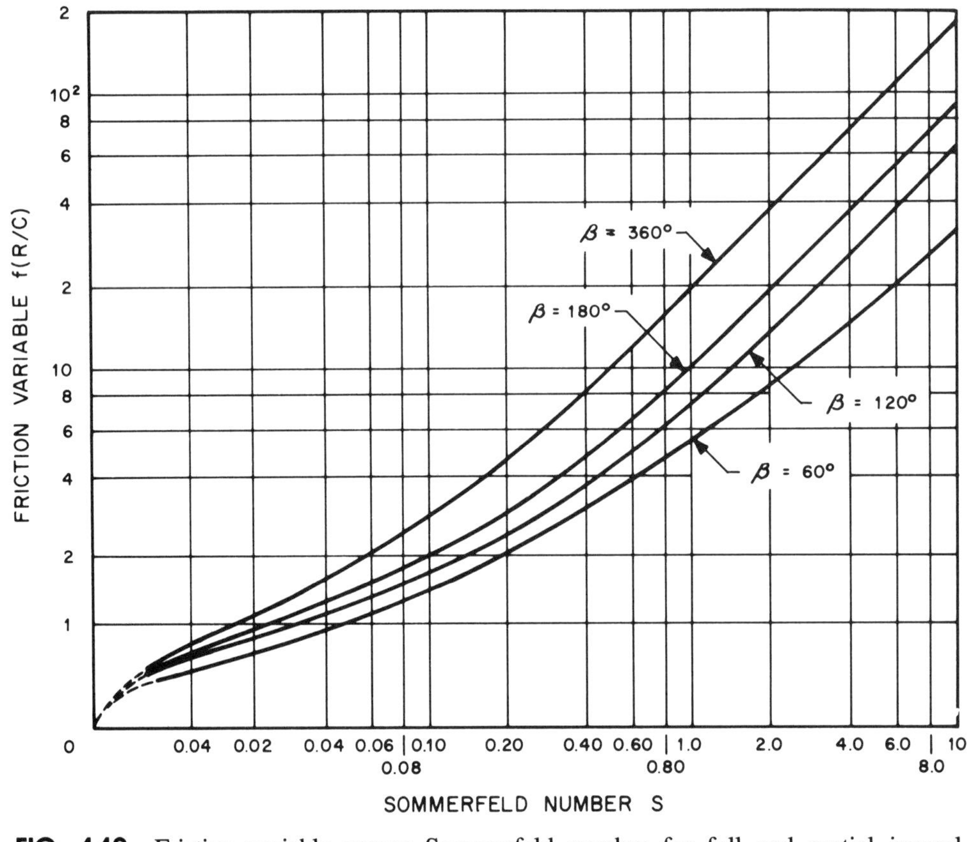

FIG. 4-10 Friction variable versus Sommerfeld number for full and partial journal bearing, $L/D = 1$, Swift-Stieber boundary conditions. *(From Raimondi and Boyd* [4-7].*)*

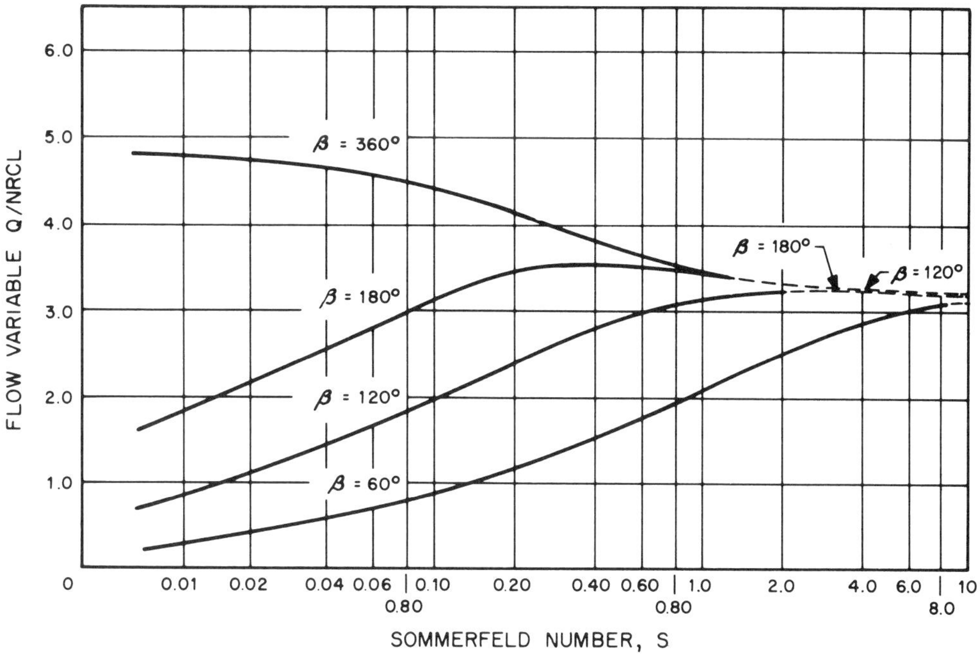

FIG. 4-11 Flow variable versus Sommerfeld number for full and partial journal bearings, $L/D = 1$, Swift-Stieber boundary conditions. *(From Raimondi and Boyd* [*4-7*].*)*

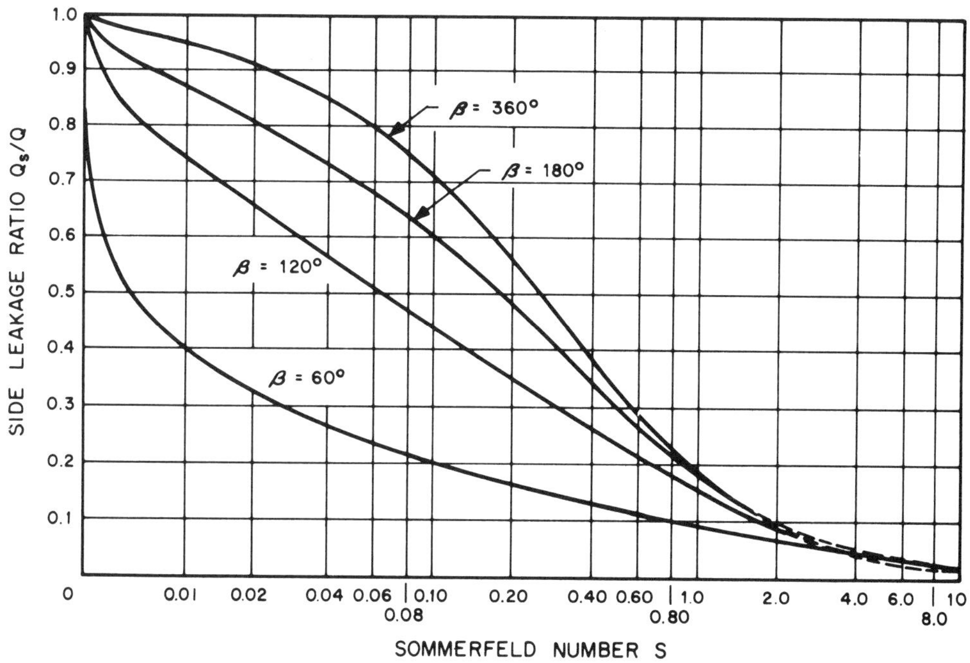

FIG. 4-12 Side-leakage ratio versus Sommerfeld number for full and partial journal bearings, $L/D = 1$, Swift-Stieber boundary conditions. *(From Raimondi and Boyd* [*4-7*].*)*

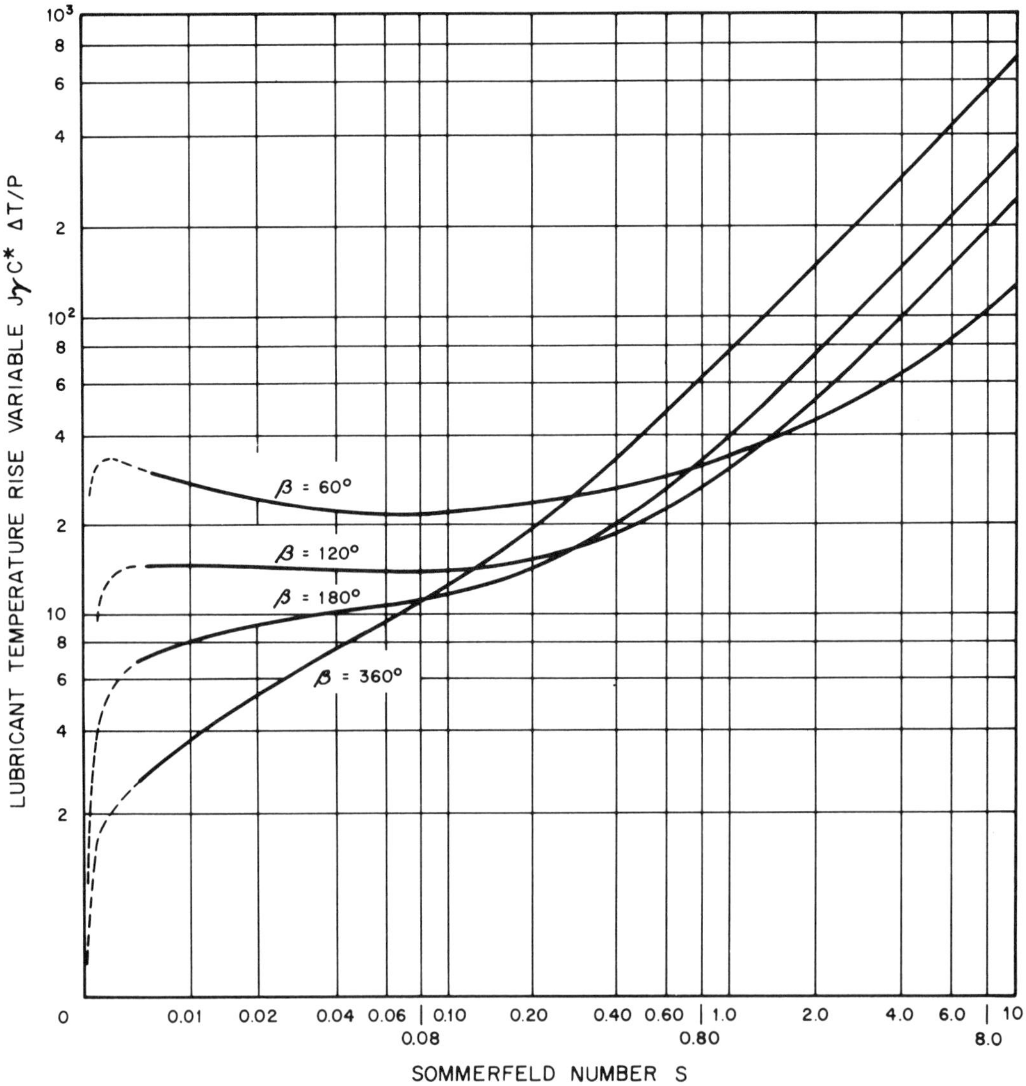

FIG. 4-13 Lubricant temperature-rise variable versus Sommerfeld number for full and partial journal bearings, $L/D = 1$, Swift-Stieber boundary conditions. *(From Raimondi and Boyd [4-7].)*

For partial bearings with bearing arc angles other than the three displayed (180°, 120°, and 60°), Raimondi and Boyd recommend using the following interpolation formula:

$$y_\beta = \frac{1}{7200}[(\beta - 120)(\beta - 60)y_{180} - 2(\beta - 180)(\beta - 60)y_{120} + (\beta - 180)(\beta - 120)y_{60}]$$

where y = any performance variable and the subscript of y is the β at which the variable is being evaluated.

Some of the tedium associated with use of charts can be removed by employing curve fits of the data. Seireg and Dandage [4-8] have developed approximate equa-

Notes ▪ Drawings ▪ Ideas

TABLE 4-14 Performance Data for Full and Partial Journal Bearings, $L/D = 1$, Swift-Stieber Boundary Conditions

L/D	ε	θ_1	S	ϕ	$\left(\frac{R}{C}\right)(f)$	$\frac{Q}{RCNL}$	$\frac{Q_s}{Q}$	$\frac{J\rho C^* \Delta T}{P}$
				Full bearing				
∞	0.1	0	0.240	69.10	4.80	3.03	0	19.9
	0.2	0	0.123	67.26	2.57	2.83	0	11.4
	0.4	0	0.0626	61.94	1.52	2.26	0	8.47
	0.6	0	0.0389	54.31	1.20	1.56	0	9.73
	0.8	0	0.0210	42.22	0.961	0.760	0	15.9
	0.9	0	0.0115	31.62	0.756	0.411	0	
1	0.1	0	1.33	79.5	26.4	3.37	0.150	106
	0.2	0	0.631	74.02	12.8	3.59	0.280	52.1
	0.4	0	0.264	63.10	5.79	3.99	0.497	24.3
	0.6	0	0.121	50.58	3.22	4.33	0.680	14.2
	0.8	0	0.0446	36.24	1.70	4.62	0.842	8.00
	0.9	0	0.0188	26.45	1.05	4.74	0.919	5.16
	0.97	0	0.00474	15.47	0.514	4.82	0.973	2.61
1/2	0.1	0	4.31	81.62	85.6	3.43	0.173	343
	0.2	0	2.03	74.94	40.9	3.72	0.318	164
	0.4	0	0.779	61.45	17.0	4.29	0.552	68.6
	0.6	0	0.319	48.14	8.10	4.85	0.730	33.0
	0.8	0	0.0923	33.31	3.26	5.41	0.874	13.4
	0.9	0	0.0313	23.66	1.60	5.69	0.939	6.66
	0.97	0	0.00609	13.75	0.610	5.88	0.980	2.56
1/4	0.1	0	16.2	82.31	322	3.45	0.180	1287
	0.2	0	7.57	75.18	153	3.76	0.330	611
	0.4	0	2.83	60.86	61.1	4.37	0.567	245
	0.6	0	1.07	46.72	26.7	4.99	0.746	107
	0.8	0	0.261	31.04	8.80	5.60	0.884	35.4
	0.9	0	0.0736	21.85	3.50	5.91	0.945	14.1
	0.97	0	0.0101	12.22	0.922	6.12	0.984	3.73
				Partial bearing, $\beta = 60°$				
∞	0.1	84.00	5.75	65.91	19.7	3.01	0	82.3
	0.2	101.00	2.66	48.91	10.1	2.73	0	46.5
	0.4	118.00	0.931	31.96	4.67	2.07	0	28.4
	0.6	126.80	0.322	23.21	2.40	1.40	0	21.5
	0.8	132.60	0.0755	17.39	1.10	0.722	0	19.2
	0.9	135.06	0.0241	14.94	0.667	0.372	0	22.5
	0.97	139.14	0.00495	10.88	0.372	0.115	0	40.7
1	0.1	82.00	8.52	67.92	29.1	3.07	0.0267	121
	0.2	99.00	3.92	50.96	14.8	2.82	0.0481	67.4
	0.4	116.00	1.34	33.99	6.61	2.22	0.0849	39.1
	0.6	125.50	0.450	24.56	3.29	1.56	0.127	28.2
	0.8	131.60	0.101	18.33	1.42	0.883	0.200	22.5
	0.9	134.67	0.0309	15.33	0.822	0.519	0.287	23.2
	0.97	139.10	0.00584	10.88	0.422	0.226	0.465	30.5

TABLE 4-14 Performance Data for Full and Partial Journal Bearings, $L/D = 1$, Swift-Stieber Boundary Conditions (*Continued*)

L/D	ε	θ_1	S	ϕ	$\left(\frac{R}{C}\right)(f)$	$\frac{Q}{RCNL}$	$\frac{Q_s}{Q}$	$\frac{J\rho C^* \Delta T}{P}$
				Partial bearing, $\beta = 60°$ (*Continued*)				
1/2	0.1	81.00	14.2	69.00	48.6	3.11	0.0488	201
	0.2	97.50	6.47	52.60	24.2	2.91	0.0883	109
	0.4	113.00	2.14	37.00	10.3	2.38	0.160	59.4
	0.6	123.00	0.695	26.98	4.93	1.74	0.236	40.3
	0.8	130.40	0.149	19.57	2.02	1.05	0.350	29.4
	0.9	134.09	0.0422	15.91	1.08	0.664	0.464	26.5
	0.97	139.22	0.00704	10.85	0.490	0.329	0.650	27.8
1/4	0.1	78.50	35.8	71.55	121	3.16	0.0666	499
	0.2	91.50	16.0	58.51	58.7	3.04	0.131	260
	0.4	109.00	5.20	41.01	24.5	2.57	0.236	136
	0.6	119.80	1.65	30.14	11.2	1.98	0.346	86.1
	0.8	128.30	0.333	21.70	4.27	1.30	0.496	54.9
	0.9	133.10	0.0844	16.87	2.01	0.894	0.620	41.0
	0.97	139.20	0.0110	10.81	0.713	0.507	0.786	29.1
				Partial bearing, $\beta = 120°$				
∞	0.1	53.300	0.877	66.69	6.02	3.02	0	25.1
	0.2	67.400	0.431	52.60	3.26	2.75	0	14.9
	0.4	81.000	0.181	39.02	1.78	2.13	0	10.5
	0.6	87.300	0.0845	32.67	1.21	1.47	0	10.3
	0.8	93.200	0.0328	26.80	0.853	0.759	0	14.1
	0.9	98.500	0.0147	21.51	0.653	0.388	0	21.2
	0.97	106.15	0.00406	13.86	0.399	0.118	0	42.4
1	0.1	47.500	2.14	72.43	14.5	3.20	0.0876	59.5
	0.2	62.000	1.01	58.25	7.44	3.11	0.157	32.6
	0.4	76.000	0.385	43.98	3.60	2.75	0.272	19.0
	0.6	84.500	0.162	35.65	2.16	2.24	0.384	15.0
	0.8	92.600	0.0531	27.42	1.27	1.57	0.535	13.9
	0.9	98.667	0.0208	21.29	0.855	1.11	0.657	14.4
	0.97	106.50	0.00498	13.49	0.461	0.694	0.812	14.0
1/2	0.1	45.000	5.42	74.99	36.6	3.29	0.124	149
	0.2	56.650	2.51	63.38	18.1	3.32	0.225	77.2
	0.4	72.000	0.914	48.07	8.20	3.15	0.386	40.5
	0.6	81.500	0.354	38.50	4.43	2.80	0.530	27.0
	0.8	92.000	0.0973	28.02	2.17	2.18	0.684	19.0
	0.9	99.000	0.0324	21.02	1.24	1.70	0.787	15.1
	0.97	107.00	0.00631	13.00	0.550	1.19	0.899	10.6
1/4	0.1	43.000	18.4	76.97	124	3.34	0.143	502
	0.2	54.000	8.45	65.97	60.4	3.44	0.260	254
	0.4	68.833	3.04	51.23	26.6	3.42	0.442	125
	0.6	79.600	1.12	40.42	13.5	3.20	0.599	75.8
	0.8	91.560	0.268	28.38	5.65	2.67	0.753	42.7
	0.9	99.400	0.0743	20.55	2.63	2.21	0.846	25.9
	0.97	108.00	0.0105	12.11	0.832	1.69	0.931	11.6

TABLE 4-14 Performance Data for Full and Partial Journal Bearings, $L/D = 1$, Swift-Stieber Boundary conditions (*Continued*)

L/D	ε	θ_1	S	ϕ	$\left(\frac{R}{C}\right)(f)$	$\frac{Q}{RCNL}$	$\frac{Q_s}{Q}$	$\frac{J\rho C^* \Delta T}{P}$
				Partial bearing, $\beta = 180°$				
∞	0.1	17.000	0.347	72.90	3.55	3.04	0	14.7
	0.2	28.600	0.179	61.32	2.01	2.80	0	8.99
	0.4	40.000	0.0898	49.99	1.29	2.20	0	7.34
	0.6	46.900	0.0523	43.15	1.06	1.52	0	8.71
	0.8	56.700	0.0253	33.35	0.859	0.767	0	14.1
	0.9	64.200	0.0128	25.57	0.681	0.380	0	22.5
	0.97	74.650	0.00384	15.43	0.416	0.119	0	44.0
1	0.1	11.500	1.40	78.50	14.1	3.34	0.139	57.0
	0.2	21.000	0.670	68.93	7.15	3.46	0.252	29.7
	0.4	34.167	0.278	58.86	3.61	3.49	0.425	16.5
	0.6	45.000	0.128	44.67	2.28	3.25	0.572	12.4
	0.8	58.000	0.0463	32.33	1.39	2.63	0.721	10.4
	0.9	66.000	0.0193	24.14	0.921	2.14	0.818	9.13
	0.97	75.584	0.00483	14.57	0.483	1.60	0.915	6.96
1/2	0.1	10.000	4.38	79.97	44.0	3.41	0.167	177
	0.2	17.800	2.06	72.14	21.6	3.64	0.302	87.8
	0.4	32.000	0.794	58.01	9.96	3.93	0.506	42.7
	0.6	45.000	0.321	45.01	5.41	3.93	0.665	25.9
	0.8	59.000	0.0921	31.29	2.54	3.56	0.806	15.0
	0.9	67.200	0.0314	22.80	1.38	3.17	0.886	9.80
	0.97	76.500	0.00625	13.63	0.581	2.62	0.951	5.30
1/4	0.1	9.000	16.3	81.40	163	3.44	0.176	653
	0.2	16.300	7.60	73.70	79.4	3.71	0.320	320
	0.4	31.000	2.84	58.99	35.1	4.11	0.534	146
	0.6	45.000	1.08	44.96	17.6	4.25	0.698	79.8
	0.8	59.300	0.263	30.43	6.88	4.07	0.837	36.5
	0.9	68.900	0.0736	21.43	2.99	3.72	0.905	18.4
	0.97	77.680	0.0104	12.28	0.877	3.29	0.961	6.46

SOURCE: Raimondi and Boyd [4-7].

tions for the full journal bearing data of the Raimondi and Boyd charts. Table 4-15 gives the coefficients to be used in these curve-fitted equations.

EXAMPLE 2. For the following data

N = 3600 rpm L = 4 in
W = 7200 lbf $C = 6.0 \times 10^3$ in
D = 6 in Lubricant: SAE 20 oil
Inlet temperature T_i = 110°F

determine the isoviscous performance of a centrally loaded full journal bearing. The viscosity-temperature relation is contained in Table 4-16.

TABLE 4-15 Seireg-Dandage Curve Fits of Raimondi-Boyd Charts: Variable $= a\left(\frac{L}{D}\right)^{b_1} S^{b_2+b_3(L/D)}$

Variable	$S \le 0.15$					$S > 0.15$				
		a	b_1	b_2	b_3		a	b_1	b_2	b_3
$\frac{1}{2} \le L/D \le 1$										
$\frac{h_0}{C} = 1 - \varepsilon$	$S \le 0.04$	2.7258	0.83621	0.75101	0.08113	$S \le 1$	0.91437	0.4538	0.6119	−0.2890
	$S > 0.04$	1.7176	1.0478	0.4999	0.1868	$S > 1$	0.89574	0.3895	0.3076	−0.2537
ϕ, rad	....	110.9067	0.60907	0.28856	0.07485	.....	74.0225	0.2395	0.3131	−0.2172
$\left(\frac{R}{C}\right)(f)$	....	9.9533	−0.4758	0.6705	−0.1124	.....	3.5251	−0.2333	−0.1926	0.1149
$\frac{Q}{RCNL}$	....	4.1036	−0.306242	−0.024799	−0.009982	.....	20.4422	−0.1125	0.8551	0.1014
$\frac{J\rho C^* \Delta T}{P}$	....	42.0097	−0.4146	0.6869	−0.1600	.....	84.2989	−0.08167	0.85540	0.08787
$\frac{p}{p_{max}}$	....	0.79567	0.59651	0.25659	0.04321	.....	0.52529	0.2486	0.2335	−0.1870

TABLE 4-15 Seireg-Dandage Curve Fits of Raimondi-Boyd Charts: Variable $= a \left(\frac{L}{D}\right)^{b_1} S^{b_2+b_3(L/D)}$ *(Continued)*

Variable	$S \le 0.15$					$S > 0.15$				
		a	b_1	b_2	b_3		a	b_1	b_2	b_3
$\frac{1}{4} \le L/D \le \frac{1}{2}$										
$\frac{h_0}{C} = 1 - \varepsilon$	$S \le 0.04$	9.2341	2.0673	0.4286	0.9247	$S \le 1$	1.1674	0.80824	0.48016	−0.02463
	$S > 0.04$	1.1545	0.4637	0.7851	−0.3788	$S > 1$	1.1263	0.7279	0.5117	−0.6581
ϕ, rad	....	112.7756	0.6332	0.2592	0.1336	.....	93.6908	0.5313	0.3139	−0.2923
$\left(\frac{R}{C}\right)(f)$	....	9.4896	−0.5446	0.7290	−0.2293	.....	17.1809	−0.3133	0.7993	0.2887
$\frac{Q}{RCNL}$	....	4.5607	−0.153864	−0.005371	−0.48838	.....	3.4980	−0.23973	−0.11107	−0.04103
$\frac{J\rho C^* \Delta T}{P}$	....	38.7604	−0.5307	0.7322	−0.2505	.....	69.4842	−0.3075	0.8003	0.2783
$\frac{p}{p_{max}}$	....	0.78635	0.57952	0.23620	0.0840	.....	0.64927	0.5037	0.2783	−0.3534

SOURCE: Ref. [4-8].

Notes ▪ Drawings ▪ Ideas

TABLE 4-16 Constants for Use in Viscosity-Temperature Equation for Various Oils

Oil	μ_0†, reyn	b, °F
SAE 10	1.58×10^{-8}	1157.5
SAE 20	1.36×10^{-8}	1271.6
SAE 30	1.41×10^{-8}	1360.9
SAE 40	1.21×10^{-8}	1474.4
SAE 50	1.70×10^{-8}	1509.6
SAE 60	1.87×10^{-8}	1564.0

†$\mu = \mu_0 \exp [b/(T + 95)]$.

SOURCE: Ref. [4-8].

Solution. Because the viscosity varies with temperature an iterative procedure is required. By this procedure a first-guess viscosity is used to determine the film temperature rise. From this an average film temperature is determined, which will permit a second film temperature rise to be determined, and so on, until a converged result is obtained.

Since $L/D = 2/3$, the Raimondi and Boyd charts would require interpolation. Alternately the Seireg-Dandage curve-fitted equations are used. The unit load may be immediately computed:

$$P = \frac{W}{LD} = \frac{7200}{(6)(4)} = 300 \text{ psi}$$

The first guess of viscosity is based on the inlet temperature:

$$\mu_1 = 1.36 \times 10^{-8} \exp \frac{1271.6}{110 + 95} = 6.72 \times 10^{-6} \text{ reyn}$$

$$S_1 = \frac{\mu_1 N}{P}\left(\frac{R}{C}\right)^2 = 0.336$$

From Table 4-15 the appropriate curve-fitted equation for the temperature rise is

$$\Delta T = 84.2989 \frac{P}{J\rho C^*}\left(\frac{L}{D}\right)^{-0.08167} S^{0.8554+0.08787L/D}$$

Taking $\rho = 0.03$ pound mass per cubic inch (lbm/in^3) and $C^* = 0.40$ Btu/(lbm·°F) as representative values for lubricating oil, we obtain

$$\Delta T_1 = 233.33(0.336)^{0.914} = 86.1°\text{F}$$

And so the second estimate of the film mean temperature is

$$T_{a2} = 110 + \frac{86.1}{2} = 153.1°\text{F}$$

Repeated calculations (13 iterations) produce

$$S = 0.176 \qquad \mu = 3.5 \times 10^{-6} \text{ reyn} \qquad \Delta T = 47.7°\text{F}$$

With the Sommerfeld number, the remaining performance parameters are easily calculated.

CONNORS' LUBRICANT SUPPLY CHARTS. The Raimondi-Boyd lubricant flow and temperature rise data are based on the notion that there is no carryover flow into the active film, that is, $Q_2 = 0$ in Eq. (4-14). From an analogous view, these results are applicable to the situation where $Q_i \gg Q_2$. Accordingly, Raimondi-Boyd predictions represent fully flooded bearing conditions and yield the coolest running lubricant temperatures for a given set of operating conditions, not accounting for any heat conduction losses.

To remedy this and thus provide more realistic design information, Connors [4-9] developed design charts which incorporate the influence of lubricant supply rate on the performance of a full journal journal bearing for $L/D = 1$. Figures 4-14 to 4-16 are plots that can be used over the entire range of flows to determine minimum

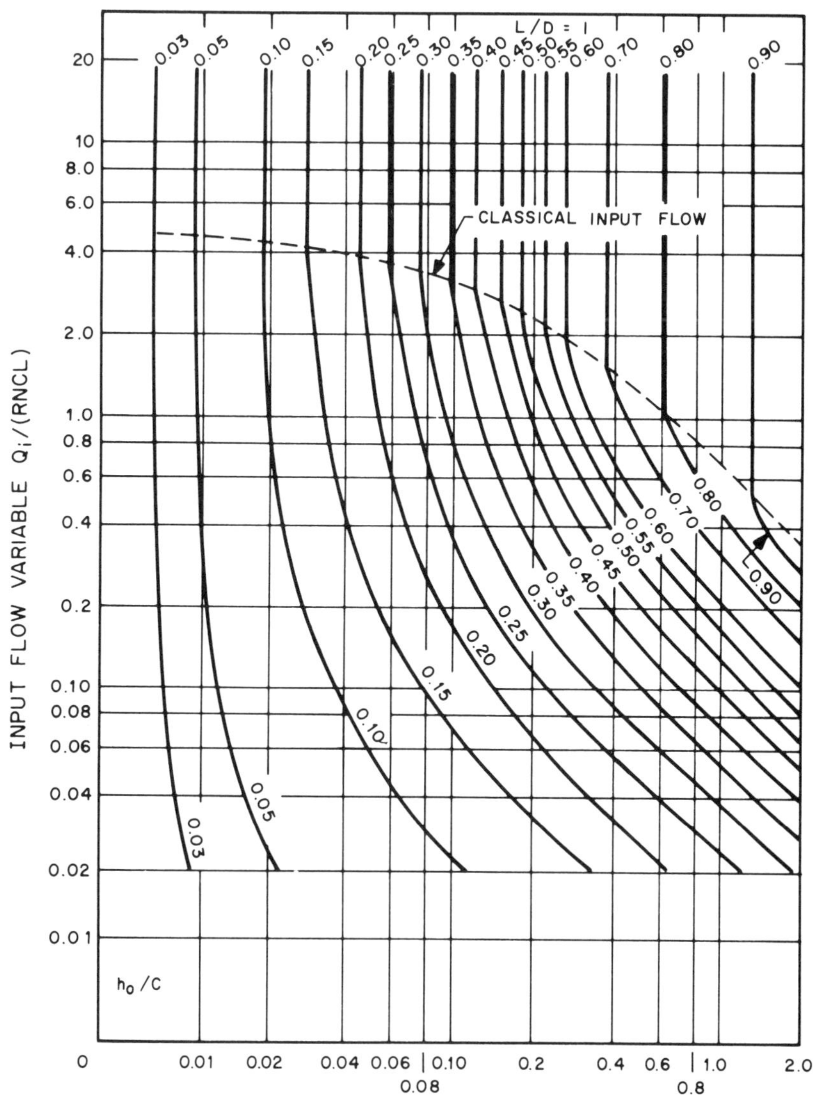

FIG. 4-14 Inlet flow variable versus Sommerfeld number for parametric values of minimum film thickness ratio; $L/D = 1$, full journal bearing. *(From Connors [4-9].)*

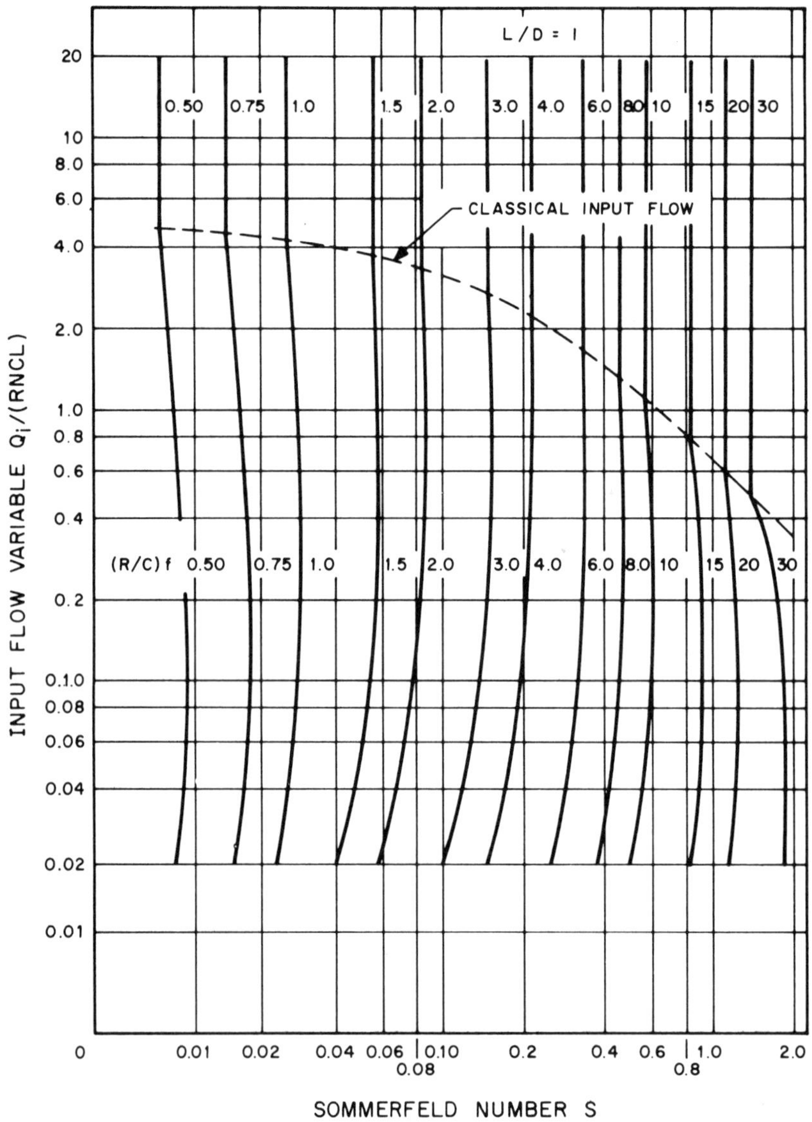

FIG. 4-15 Inlet flow variable versus Sommerfeld number for parametric values of friction variable; $L/D = 1$, full journal bearing. *(From Connors [4-9].)*

film thickness, friction, and temperature rise from given values of the Sommerfeld number and the inlet flow variable.

EXAMPLE 3. Determine the lubricant temperature rise as a function of the inlet flow rate for the following design parameters:

$W = 1500$ lbf	$C = 4 \times 10^{-3}$ in
$N = 1800$ rpm	SAE 30 oil
$D = 4$ in	$T_i = 100°F$
$L = 4$ in	

Notes ▪ Drawings ▪ Ideas

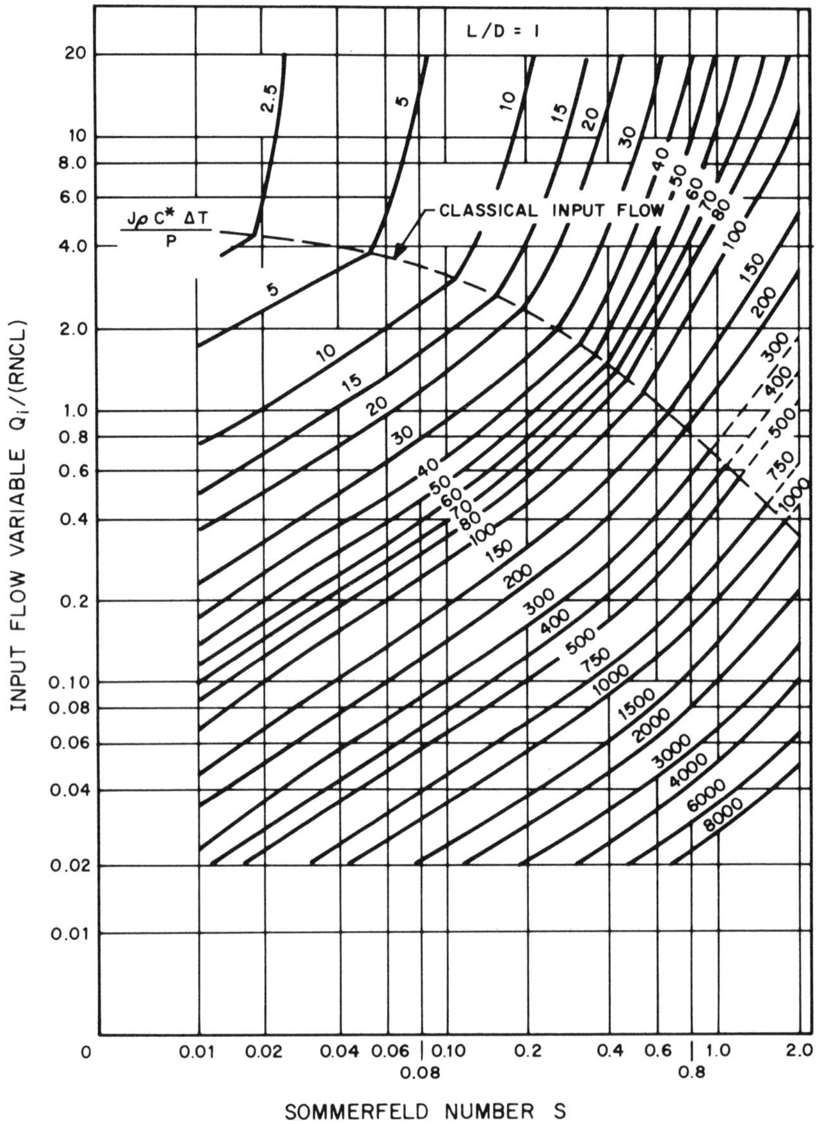

FIG. 4-16 Inlet flow variable versus Sommerfeld number for parametric values of temperature-rise variable; $L/D = 1$, full journal bearing. *(From Connors* [*4-9*].*)*

Solution. To solve this type of problem, a plot or equation of lubricant viscosity as a function of temperature must be available. The calculation procedure is as follows:

1. Select a value of $Q_i/(RCNL)$.
2. Assume a viscosity value.
3. Compute the Sommerfeld number.
4. Use the $Q_i/(RCNL)$ and S values to find $J\rho C^*(T_a - T_i)/P$ in Fig. 4-16.
5. Calculate the mean film temperature T_a.
6. Increment μ and repeat the process from step 3 until there are sufficient points to

Notes ▪ Drawings ▪ Ideas

establish an intersection with the lubricant's μ versus T data. This intersection represents the operating point for the given $Q_i/(RNCL)$.

7. Increment the input flow variable, and return to step 2.

For $Q_i/(RCNL) = 1$, the following sets of data were obtained by performing this calculation procedure:

$$\mu = \begin{cases} 1.5 \times 10^{-6} \text{ reyn} \\ 3.0 \times 10^{-6} \text{ reyn} \\ 6.0 \times 10^{-6} \text{ reyn} \end{cases} \qquad S = \begin{cases} 0.12 \\ 0.24 \\ 0.48 \end{cases}$$

$$\frac{J\rho C^*(T_a - T_i)}{P} = \begin{cases} 32 \\ 60 \\ 115 \end{cases} \qquad T_a = \begin{cases} 126.8°\text{F} \\ 150.2°\text{F} \\ 196.2°\text{F} \end{cases}$$

Using these and the lubricant μ versus T relation as presented in Table 4-16, we find the operating point to be

$$T_a = 155°\text{F} \qquad \mu = 3.2 \times 10^{-6} \text{ reyn}$$

Hence, $S = 0.256$. Also from Fig. 4-14 we obtain $h_0/C = 0.52$, so $h_0 = 0.00208$. Further, from Fig. 4-15, $(R/C)(f) = 5$ so $f = 0.01$, which allows us to calculate the power loss to be 0.857 horsepower (hp). Assuming other values of $Q_i/(RCNL)$ permits Fig. 4-17 to be drawn. The Raimondi-Boyd value corresponding to $Q_i \rightarrow \infty$ is also presented.

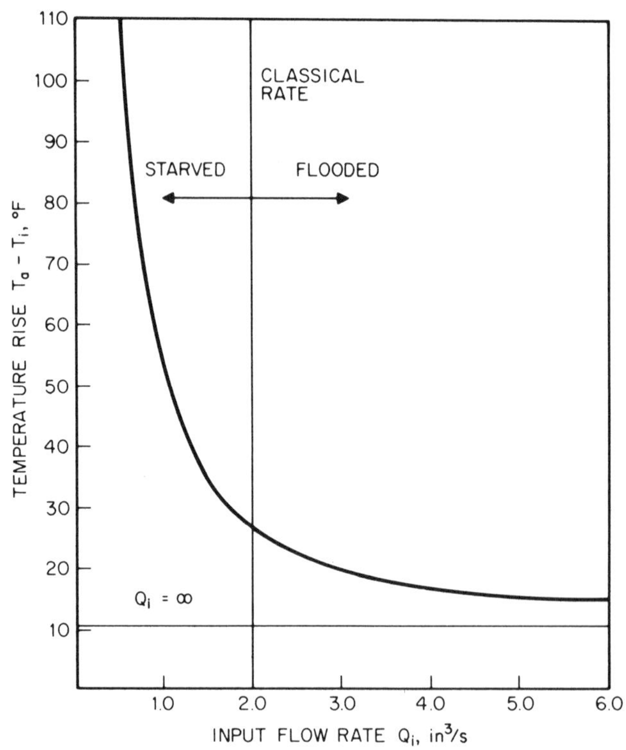

FIG. 4-17 Lubricant temperature rise versus lubricant input flow rate (Example 3).

In Sec. 4-5-4 it was shown that

$$(T_a - T_i)_{\text{conduction}} = (1 - \lambda)(T_a - T_i)_{\text{no conduction}}$$

where λ = ratio of heat conduction to heat generation rate and is assumed to be a constant. By using this idea a new operating point for a given $Q_i/(RCNL)$ can be determined. For example, with $\lambda = 0.25$ and $Q_i/(RCNL) = 1$, we find that T_a = 147°F, $h_0 = 0.0023$, and HP = 0.960 hp.

4-6-3 Optimization

In designing a journal bearing, a choice must be made among several potential designs for the particular application. Thus the designer must establish an optimum design criterion for the bearing. The design criterion describes the designer's objective, and numerous criteria can be envisioned (e.g., minimizing frictional loss, minimizing the lubricant temperature rise, minimizing the lubricant supply to the bearing, and so forth).

The search for an optimum bearing design is best conducted with the aid of a computer. However, optimum bearing design can also be achieved graphically. Moes and Bosma [4-10] developed a design chart for the full journal bearing which enables the designer to select optimum bearing dimensions. This chart is constructed in terms of two dimensionless groups called X and Y here. The groups include two quantities of primary importance to the bearing designer: minimum film thickness h_0 and frictional torque M_j; the groups do not contain the bearing clearance. The dimensionless groups are

$$X \equiv \frac{h_0}{R}\left(\frac{P}{2\pi N\mu}\right)^{1/2} \qquad Y \equiv \frac{M_j}{WR}\left(\frac{P}{2\pi N\mu}\right)^{1/2} \tag{4-15}$$

Both X and Y can be written in terms of the Sommerfeld number. Recalling that $h_0 = C(1 - \varepsilon)$ and $S = (\mu N/P)(R/C)^2$, we can easily show that

$$X = \frac{1 - \varepsilon}{\sqrt{2\pi S}} \qquad \text{and} \qquad Y = \frac{M_j}{WC}\frac{1}{\sqrt{2\pi S}}$$

Figure 4-18 is a plot of full journal bearing design data on the XY plane. In the diagram two families of curves can be distinguished: curves of constant L/D ratio and curves of constant ε. Use of this diagram permits rather complicated optimization procedures to be performed.

EXAMPLE 4. Calculate the permissible range of minimum film thickness and bearing clearance that will produce minimum shaft torque for a full journal bearing operating under the following conditions:

$\mu = 5 \times 10^{-6}$ reyn $\quad D = 4$ in
$N = 1800$ rpm $\quad L = 3$ in
$W = 1800$ lbf

Solution. As a first step, we calculate the largest h_0 for the given conditions. This is easily accomplished by locating the coordinates on Fig. 4-18 corresponding to the maximum X for $L/D = 3/4$, or

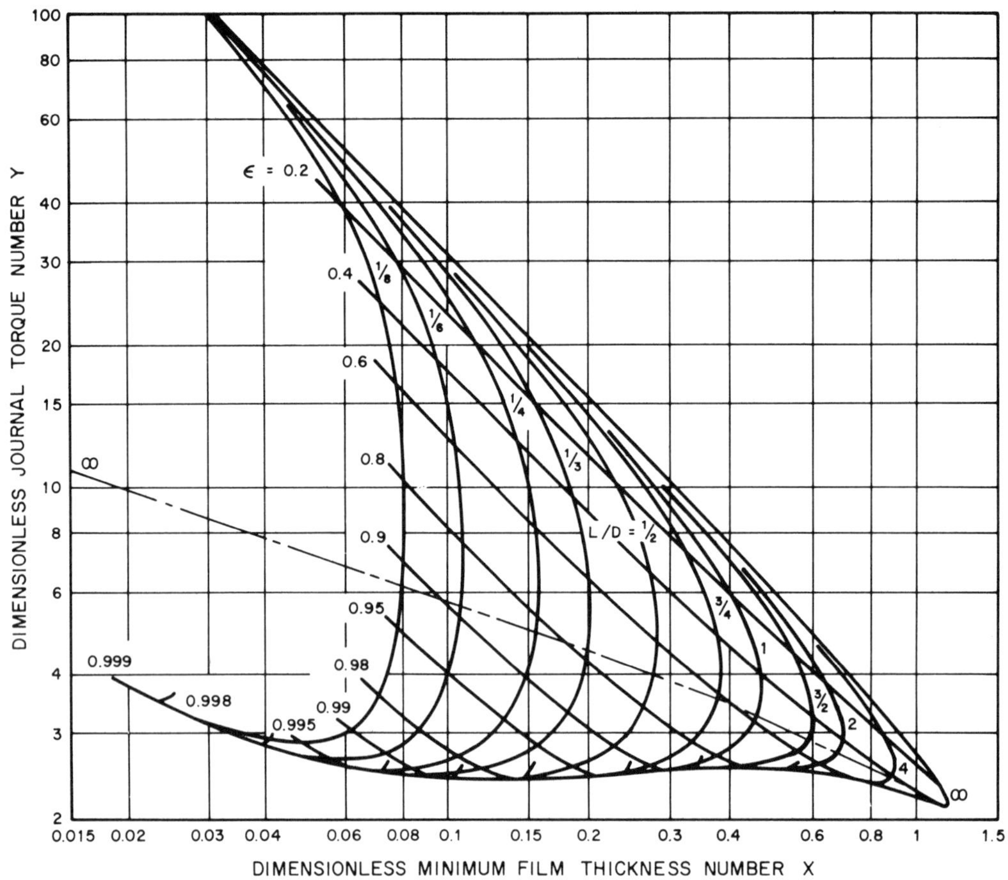

FIG. 4-18 Optimization chart for full journal bearings. *(From Moes and Bosma [4-10].)*

$$\begin{Bmatrix} X \\ Y \\ \varepsilon \end{Bmatrix} = \begin{Bmatrix} 0.385 \\ 4.0 \\ 0.54 \end{Bmatrix} \qquad \text{at } X = X_{\max}$$

Thus, from Eq. (4-15),

$$h_0 = 1.93 \times 10^{-3} \text{ in} \qquad \text{and} \qquad M_j = 36.1 \text{ in} \cdot \text{lbf}$$

The clearance is calculated from $C = h_0/(1 - \varepsilon) = 4.2 \times 10^{-3}$ in.

Next the coordinates on Fig. 4-18 corresponding to the minimum shaft torque (minimum value of Y at $L/D = 3/4$) are located at

$$\begin{Bmatrix} X \\ Y \\ \varepsilon \end{Bmatrix} = \begin{Bmatrix} 0.18 \\ 2.5 \\ 0.97 \end{Bmatrix} \qquad \text{at } Y = Y_{\min}$$

Thus, from Eq. (4-15),

$$h_0 = 9.02 \times 10^{-4} \text{ in} \qquad \text{and} \qquad M_j = 22.6 \text{ in} \cdot \text{lbf}$$

The clearance for this case is calculated to be 3.01×10^{-2} in.

Notes ▪ Drawings ▪ Ideas

Thus the two optima (largest film thickness and smallest shaft torque) do not coincide. Thus, a compromise clearance value must generally be selected. Figure 4-19 is a plot of recommended minimum clearance for given shaft speed and diameter, which can be used as a helpful guide in this process.

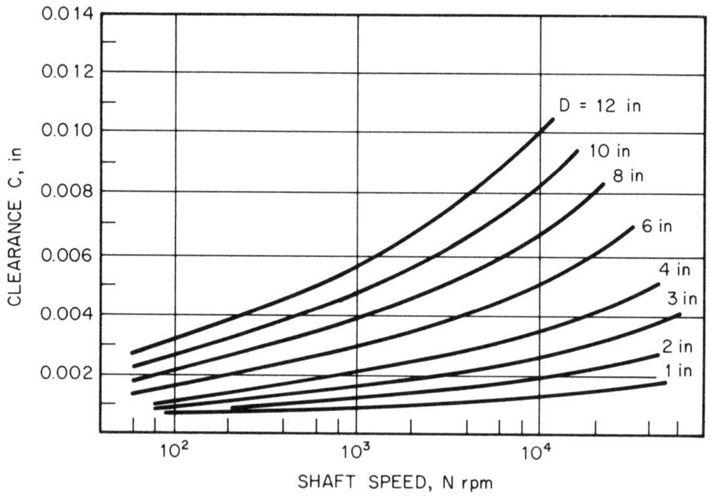

FIG. 4-19 Recommended minimum clearance versus journal speed for a given journal diameter.

EXAMPLE 5. Determine the bearing length and clearance for a full journal bearing operating steadily with minimum friction loss for the following conditions:

$\mu = 17.83 \times 10^{-6}$ $\quad W = 1500$ lbf
$N = 4800$ rpm $\quad D = 6$ in

and a minimum allowable film thickness of

$$(h_0)_{\min} = 4.4 \times 10^{-4}$$

Solution. Because both X and Y contain the bearing length L, a new set of dimensionless groups not containing L must be developed. This pair is easily seen to be

$$U = (X)\left(\frac{L}{D}\right)^{1/2} = \frac{h_0}{R}\left(\frac{W}{2\pi\mu ND^2}\right)^{1/2} \tag{4-16}$$

$$V = (Y)\left(\frac{L}{D}\right)^{1/2} = \frac{M_j}{WR}\left(\frac{W}{2\pi\mu ND^2}\right)^{1/2} \tag{4-17}$$

For the given information the smallest value of U (termed $U_{\min}$ because the smallest value of h_0 is used) can be directly calculated from Eq. (4-16):

$$U_{\min} = 0.10 \qquad \text{Thus,} \qquad X \geq (0.10)\left(\frac{D}{L}\right)^{1/2}$$

Selected L/D values permit an array of X values to be calculated. This in turn permits the minimum values of Y (that is, $Y_{\min}$) to be read from Fig. 4-18. Then cor-

Notes ▪ Drawings ▪ Ideas

responding V_{min} values can be calculated from Eq. (4-17). Table 4-17 contains the results of this operation.

TABLE 4-17 Summary of Calculated Results for Example 5

L/D	X	Y_{min}	V_{min}
1/4	0.2	No solution	
1/3	0.173	3.20	1.85
1/2	0.414	2.43	1.72
3/4	0.115	2.40	2.08
1	0.100	2.40	2.40

Next a cubic equation is fitted to the V_{min} data, and we find

$$V_{min} = 4.057 - 11.927\frac{L}{D} + 18.742\left(\frac{L}{D}\right)^2 - 8.472\left(\frac{L}{D}\right)^3 \tag{4-18}$$

The optimum value of L/D for a minimum friction loss can be found by differentiation, which yields the optimum bearing length:

$$L = 0.464D = 2.78 \text{ in}$$

The frictional loss can be calculated by inserting the optimum L/D ratio in the cubic equation:

$$V_{min} = (Y)\left(\frac{L}{D}\right)^{1/2} = 1.712$$

Thus, $Y = 2.513$, and the eccentricity ratio is read from Fig. 4-18 as $\varepsilon \cong 0.98$. Finally the optimum bearing clearance is calculated to be $C = 2.2 \times 10^{-2}$ in.

4-7 GAS-LUBRICATED JOURNAL BEARINGS

Gas-lubricated journal bearings have been employed in a wide variety of modern industrial applications. For example, they are used in dental drills, high-speed machine tools, digital-computer peripheral devices, high-speed turbomachines, and navigational instruments.

Gas bearings produce very little friction even at high speeds; hence, they have low frictional losses and generate minimal amounts of heat. In addition, gas lubricants are chemically very stable over a wide range of temperatures; they neither freeze nor boil; they are nonflammable; and they do not contaminate bearing surfaces. Gas bearings also have low noise characteristics.

On the negative side, gas bearings do not have much load-carrying capacity and have large startup wear. Also, because the gas film thickness is quite small, gas bearings require superior surface finish and manufacture. Great care must be exercised with the journal alignment. Further, thin films offer very little cushion or damping capacity; consequently, gas bearings are prone to certain vibrational instabilities.

4-7-1 Limiting Gas Bearing Solutions

The compressible-flow Reynolds equation, Eq. (4-7), is too complex to permit a general analytical solution, and numerical methods have been used to obtain gas bearing information. However, considerable insight can be gained by considering solutions for limiting cases. Besides the limiting geometric problems of long and short bearings, two other limiting classes of problems are important: gas films with low bearing numbers and gas films with high bearing numbers. The former pertain to low-speed, high-loading applications while the latter pertain to very-high-speed, light-loading applications. Some bearing performance parameters for gas films with low and high bearing numbers are shown in Table 4-18.

TABLE 4-18 Long Gas Journal Bearing Performance Parameters, Limiting Cases, Long-Bearing Theory

Performance parameter	Low bearing number	Large bearing number
$\frac{p}{p_a}$	$\frac{1 + \Lambda\varepsilon \sin\theta(2 + \varepsilon\cos\theta)}{(2 + \varepsilon)^2(1 + \varepsilon\cos\theta)^2}$	$\frac{\sqrt{1 + 3\varepsilon^2/2}}{1 + \varepsilon\cos\theta}$
$\frac{W_R}{p_a DL}$	0	$\frac{\pi}{2}\left(1 + \frac{3\varepsilon^2}{2}\right)^{1/2}[(1 - \varepsilon^2)^{-1/2} - 1]$
$\frac{W_T}{p_a DL}$	$\frac{\pi\varepsilon\Lambda}{(2 + \varepsilon^2)\sqrt{1 - \varepsilon^2}}$	0
ϕ	$\frac{\pi}{2}$	0
$\frac{F_j}{\mu UL}\frac{C}{R}$	$\frac{4\pi(1 + 2\varepsilon^2)}{(2 + \varepsilon^2)\sqrt{1 - \varepsilon^2}}$	$\frac{2\pi}{\sqrt{1 - \varepsilon^2}}$
$\left(\frac{R}{C}\right)(f)$	$\frac{1 + 2\varepsilon^2}{3\varepsilon}$	$\frac{2\mu UR}{C^2 P_a\sqrt{(1 + 3\varepsilon^2/2}(1 - \sqrt{1 - \varepsilon^2})}$

4-7-2 Stability Considerations

The journal bearing of a rotating machine cannot be treated separately. It is part of a complex dynamic system, and its design can influence the dynamic behavior of the entire system. Bearing damping controls vibration amplitude and tolerance to any imbalance and, to a large extent, determines whether the rotor will be dynamically stable. A bearing running in an unstable mode can lose its ability to support load, and rubbing contact can occur between the bearing and shaft surfaces.

Rotor bearing instabilities are particularly troublesome in gas-lubricated bearings; the film is considerably thinner in the gas bearing than in the liquid-lubricated bearing and so does not possess the damping capacity. We can differentiate between two forms of dynamic instability: *synchronous whirl* and *half-frequency whirl* (also called fractional frequency whirl and film whirl).

SYNCHRONOUS WHIRL. A rotating shaft experiences periodic deflection (forced vibration) because of the distribution of load, the method of shaft support, the degree of flexibility of the shaft, and any imbalance within the rotating mass. This deflection causes the journal to orbit within its bearings at the rotational speed. When the frequency of this vibration occurs at the natural frequency (or critical speed) of the system, a resonance condition exists. In this condition, the amplitude of vibration (size of the journal orbit) increases and can cause bearing failure. Because the shaft rotational speed and the critical speed coincide, this form of instability is termed *synchronous whirl.*

Since stable operation occurs on either side of the critical speed, the bearing system must be designed so that critical speeds do not exist in the operating speed range. This can be accomplished by making the critical speeds either very large or very small. Large critical speeds can be established by increasing the bearing stiffness by reducing the bearing clearance. Lower critical speeds can be established by increasing the shaft flexibility. The instability can also be suppressed by mounting the bearing in a flexible housing which introduces additional system damping.

HALF-FREQUENCY WHIRL. A deflected rotating shaft has its center whirl, or orbit, around the center of the bearing. When the whirl speed Ω is equal to or somewhat less than half the rotational speed of the shaft ω, the hydrodynamic capacity of the bearing to support a load is diminished and, in fact, may fall to zero. With any subsequent rotational speed increase, the whirl amplitude will increase and the bearing and journal may come into violent contact. This form of self-activated vibrational instability is called *half-frequency,* or *half-speed whirl.* Because the whirl speed for gas bearings is generally less than the rotational speed, the instability is also termed *fractional frequency whirl.* Figure 4-20 is a map for assessing the stability of a full journal bearing.

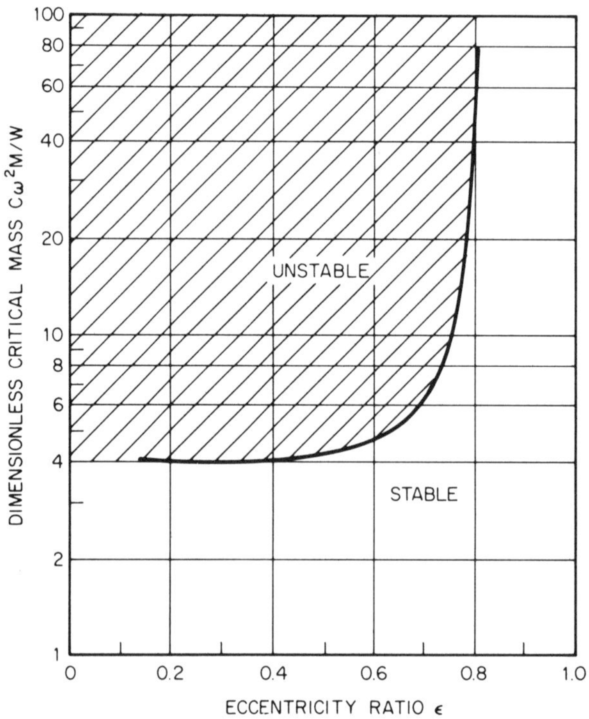

FIG. 4-20 Bearing stability for full journal bearing, $L/D = 1$.

Half-speed whirl may be suppressed by altering the circumferential symmetry of the bearing. This can be accomplished by providing axial grooves in the bearing surface by dividing the circumference into a number of sections, or lobes, or by using tilting pads. Also, introducing more damping into the system through the use of flexible housing mounts can be helpful.

4-7-3 Cylindrical Ungrooved Gas Journal Bearings

Complete plain cylindrical journal bearings are generally used in applications that require support of a fixed, unidirectional load. Self-excited whirl instability is a common problem with high-speed rotor systems supported by this type of bearing. However, if the load is sufficiently large to have an eccentricity ratio on the order of 0.8, whirl problems are reduced (refer to Fig. 4-20).

Ungrooved cylindrical gas bearings have been analyzed by Raimondi [4-11]. Performance characteristics were developed numerically and presented in the form of various design charts for L/D ratios of 1/2, 1, and 2. Figures 4-21 to 4-23 are a sampling of these charts for an L/D ratio of 1.

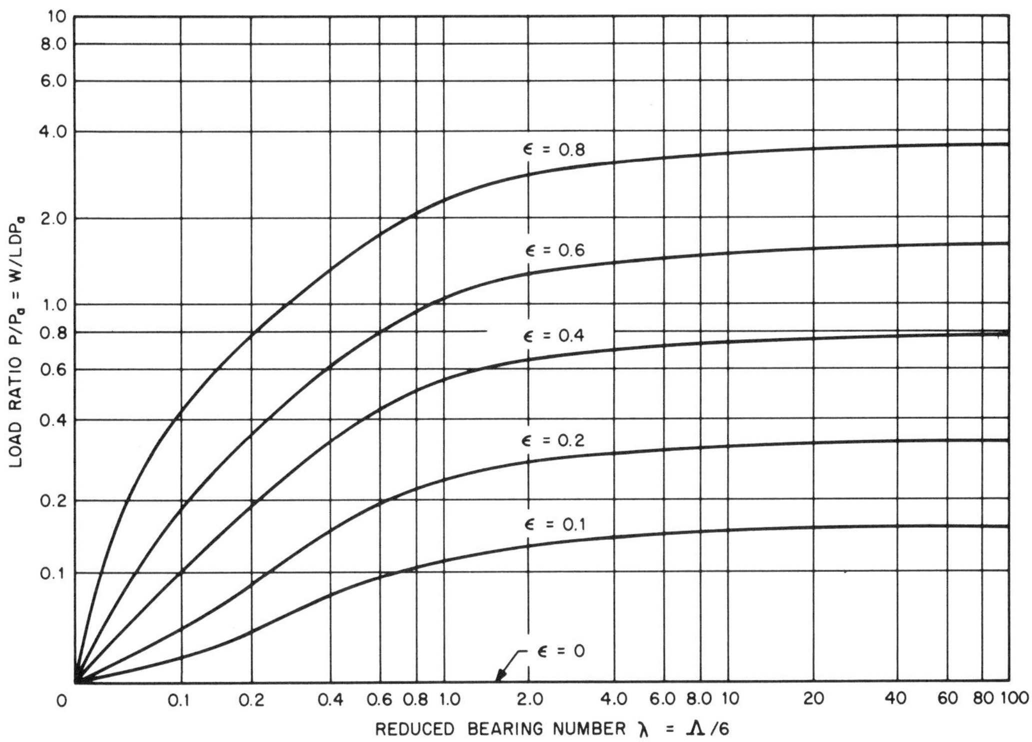

FIG. 4-21 Load ratio versus reduced bearing number of an ungrooved cylindrical gas bearing; $L/D = 1$. *(From Raimondi [4-11].)*

A porous journal bearing is a plain cylindrical journal bearing with a porous liner that is fixed in the bearing housing (Fig. 4-24). An externally pressurized gas is supplied to the outer surface of the liner and flows through the porous material into the bearing clearance space.

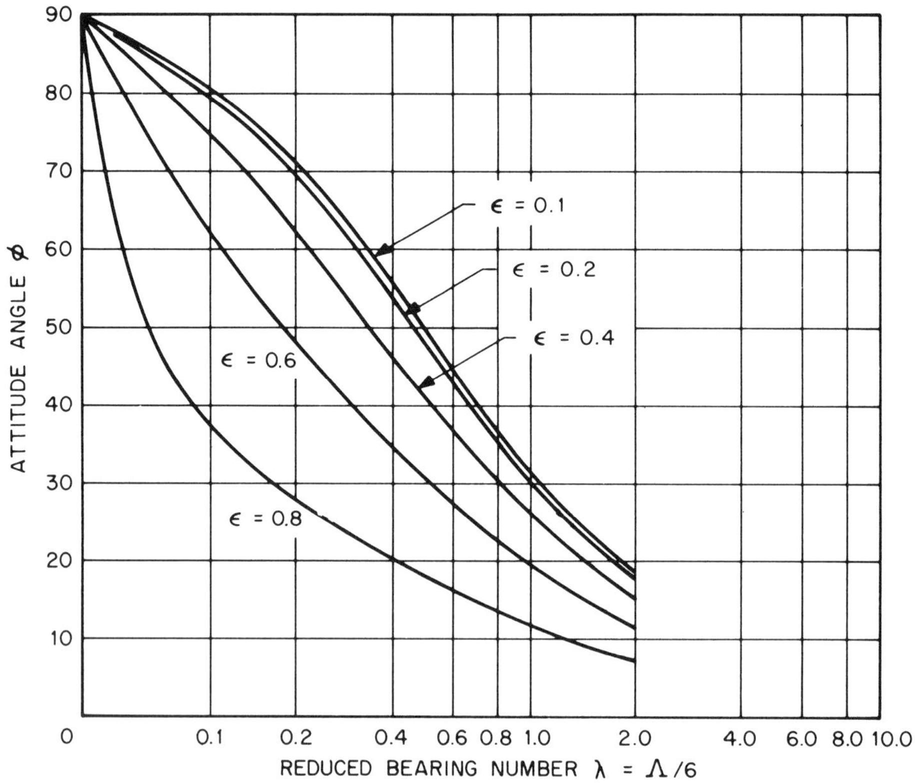

FIG. 4-22 Attitude angle versus reduced bearing number of an ungrooved cylindrical gas bearing; $L/D = 1$. *(From Raimondi [4-11].)*

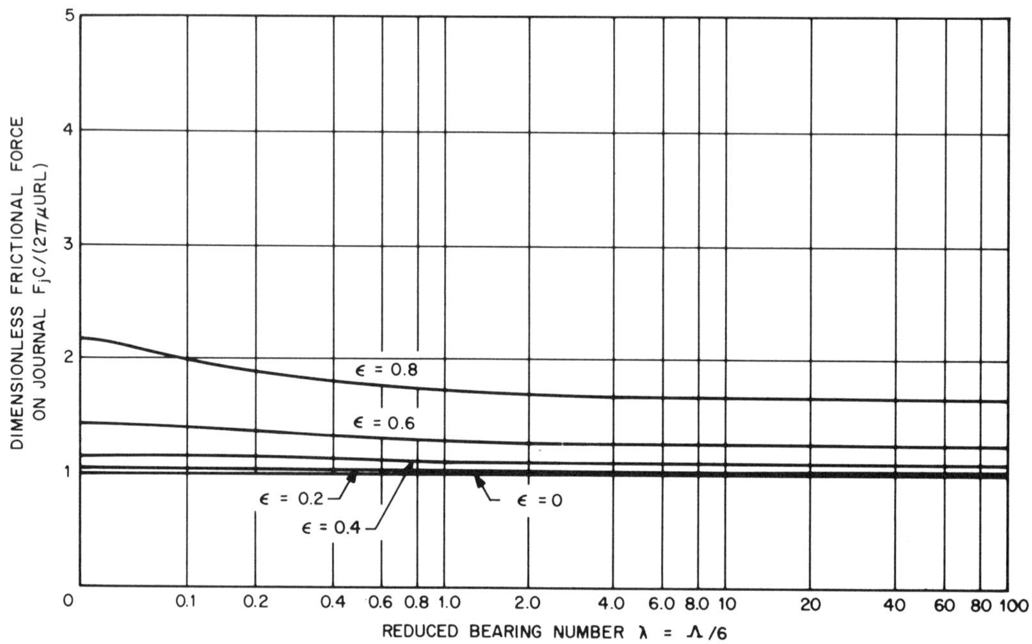

FIG. 4-23 Dimensionless journal frictional force versus reduced bearing number of an ungrooved cylindrical gas bearing; $L/D = 1$. *(From Raimondi [4-11].)*

Notes ▪ Drawings ▪ Ideas

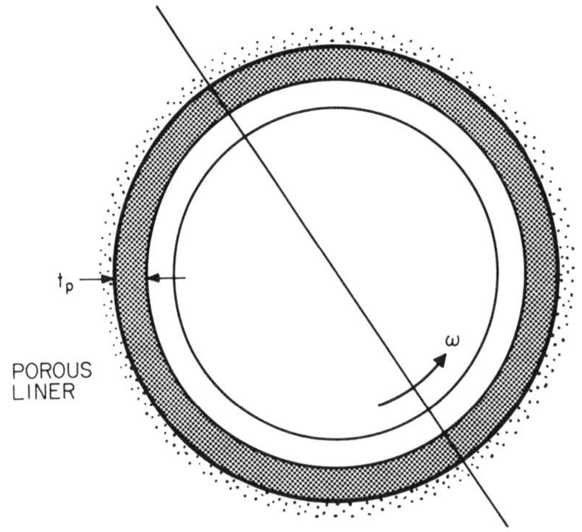

FIG. 4-24 Porous journal bearing.

The steady performance of a self-acting porous gas bearing of finite length has been determined by Wu [4-12]. Table 4-19 shows a sample of these performance data. The data are applicable for a particular porous linear thickness ratio (0.083) and for particular combinations of the *slip coefficient* α (dependent on the structure of the porous material), the *permeability* k (a physical property of the porous material), and bearing dimensions R and C.

EXAMPLE 6. For the following conditions

$D = 0.5$ in	$N = 47\,000$ rpm
$L = 0.5$ in	$\mu = 2.6 \times 10^{-9}$ reyn (air at 80°F)
$C = 2 \times 10^{-4}$ in	$p_a = 40$ psi
	$W = 6.1$ lbf

compare the performance of a porous gas bearing to that of a solid wall bearing. The porous bearing material is ceramic 0.021 in thick with a permeability $k = 1.33 \times 10^{-12}$ in^2 and a slip coefficient $\alpha = 0.3$.

Solution. Since $L/D = 1$, $t_p/R \cong 0.083$, $\beta = \alpha C/k \cong 50$, and $\Omega = 12kR/C^3 \cong 0.5$, the data of Table 4-19 may be used. The bearing number is

$$\Lambda = \frac{6\mu\omega}{p_a}\left(\frac{R}{C}\right)^2 = 3.0$$

The load ratio is

$$\frac{P}{p_a} = \frac{W/(LD)}{p_a} = 0.61$$

Entering Table 4-18 with these values, we find

$$\varepsilon = 0.6 \qquad \phi = 37.598 \qquad \left(\frac{R}{C}\right)(f) = 3.3003$$

For the same values we can find from Figs. 4-21 to 4-23 for the solid wall bearing $\varepsilon = 0.54$, $\phi = 33.7°$, and

TABLE 4-19 Gas-Lubricated Bearing Performance Data, $L/D = 1$, $t_p/R = 0.083$, $\Omega = 0.5$, $\beta = 50$

ε	Λ	$\frac{P}{p_a}$	ϕ, deg	$\left(\frac{R}{C}\right)(f)$
0.2	0.6	0.039	80.563	8.1034
	1.2	0.0756	72.109	8.3784
	3.0	0.1594	52.470	9.9199
	6.0	0.2272	34.402	13.8773
	12.0	0.2662	20.290	23.6302
	50.0	0.2959	7.602	88.4725
	100.0	0.3054	5.014	171.9223
	∞	0.3264	0	∞
0.4	0.6	0.0853	77.245	4.0835
	1.2	0.1638	66.670	4.2517
	3.0	0.3441	46.382	5.0012
	6.0	0.5067	30.262	6.7025
	12.0	0.6164	17.806	10.9197
	50.0	0.6982	6.424	39.9776
	100.0	0.7188	4.164	77.6340
	∞	0.7746	0	∞
0.6	0.6	0.1500	70.129	2.8072
	1.2	0.2872	56.903	2.8961
	3.0	0.6102	37.598	3.3003
	6.0	0.9322	24.537	4.2078
	12.0	1.1854	14.488	6.4992
	50.0	1.3774	4.962	23.0667
	100.0	1.4186	3.122	44.7611
	∞	1.5603	0	∞
0.8	0.6	0.2547	61.657	2.2908
	1.2	0.4995	47.468	2.2971
	3.0	1.1050	31.147	2.4980
	6.0	1.7837	20.348	2.9210
	12.0	2.3907	12.272	4.2374
	50.0	2.8736	4.046	14.4249
	100.0	2.9640	2.436	27.9331
	∞	3.4193	0	∞

SOURCE: Ref. [4-12].

$$\frac{F_j C}{2\pi\mu URL} = 1.18$$

From the last value we may compute for comparative purposes

$$\left(\frac{R}{C}\right)(f) = 3.038$$

Thus the porous bearing operates at a larger eccentricity ratio (which indicates better stability) but has larger frictional loss. On the other hand, for a given ε, the porous bearing has a lower load capacity compared to the solid wall bearing. This loss of load capacity becomes more severe as the eccentricity ratio is increased.

4-7-4 Axially Grooved Gas Journal Bearings

The addition of axial slots, or grooves, in the bearing surface can produce several positive effects. Feed grooves can reduce frictional heating and energy losses. The grooves, in effect, turn the surface into a number of partial arc bearings, which can reduce the tendency for half-frequency whirl instability. What is more, there is only a slight increase in the manufacturing costs for grooving a bearing. Although a variety of grooving arrangements can be envisioned, three or four equally spaced axial grooves are typically employed. Feed grooves of 30° to 60° extent are commonly used. However, bearing stability decreases with increase in the groove size.

Castelli and Pirvics [4-13] presented tabular design data for axially grooved gas bearings. The subtended angle of each groove was taken to be 5°. Figures 4-25 and

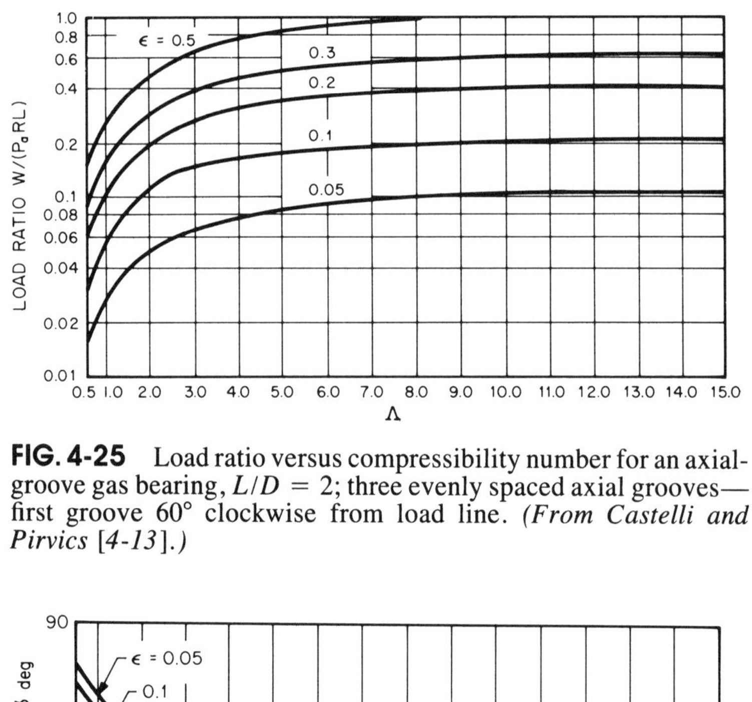

FIG. 4-25 Load ratio versus compressibility number for an axial-groove gas bearing, $L/D = 2$; three evenly spaced axial grooves—first groove 60° clockwise from load line. *(From Castelli and Pirvics [4-13].)*

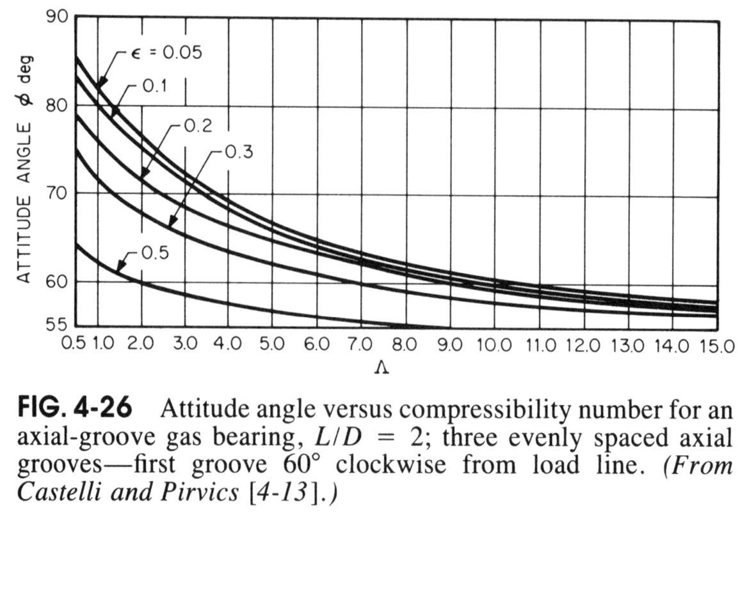

FIG. 4-26 Attitude angle versus compressibility number for an axial-groove gas bearing, $L/D = 2$; three evenly spaced axial grooves—first groove 60° clockwise from load line. *(From Castelli and Pirvics [4-13].)*

Notes ▪ Drawings ▪ Ideas

TABLE 4-20 Eccentricity Ratio versus Sommerfeld and Bearing Numbers for Circular, Elliptical, and Three-Lobe Noncircular Gas Bearings

		Eccentricity ratio ε				
Bearing number Λ	Sommerfeld number S	Circular	Elliptical central loading	Elliptical optimum loading	Three-lobe central loading	Three-lobe optimum loading
1	0.365	0.2	0.54	0.525	0.55	0.545
1	0.162	0.4	0.59	0.56	0.615	0.605
1	0.0866	0.6	0.67	0.635	0.70	0.685
1	0.0381	0.8	0.815	0.77	0.84	0.815
3	0.365	0.25	0.57	0.535	0.595	0.57
3	0.162	0.475	0.65	0.59	0.68	0.645
3	0.0860	0.655	0.75	0.685	0.80	0.755

4-26 are sample design chart data for a gas bearing with three grooves and an L/D ratio of 2.

4-7-5 Noncircular Gas Journal Bearings

For low eccentricity ratios, a noncircular gas bearing has superior stability properties but inferior load capacity when it is compared with a circular gas bearing. At higher eccentricity ratios, the load-carrying capacity of the noncircular gas bearing exceeds that of the circular gas bearing. For applications where load capacity is not an issue (e.g., in vertical-shaft cases), use of a noncircular gas bearing minimizes stability concerns that exist at the lower eccentricity ratios.

Pinkus [4-14] has investigated the performance of elliptical and three-lobe gas bearings. It was found that the direction of load application is important to the performance of a noncircular gas bearing. In fact, load-carrying capacity can be optimized by rotating the bearing relative to the load line. The rotation is generally clockwise anywhere from a few degrees to 25°. Table 4-20 presents a comparison of expected eccentricity ratios for given values of S and Λ for circular, elliptical, and three-lobe gas bearings. Higher values of ε required by the noncircular bearings over the ε for the circular bearing at the same Sommerfeld number indicate a lower load capacity. Pinkus also presented envelopes of operation. Figure 4-27 is a plot of Sommerfeld number versus the eccentricity ratio for a noncentrally loaded elliptical gas bearing. The upper set of curves represents the largest ε for a given S at a particular Λ, and it would be used in attempt to avoid stability problems. The lower set of curves represents the optimum loading conditions.

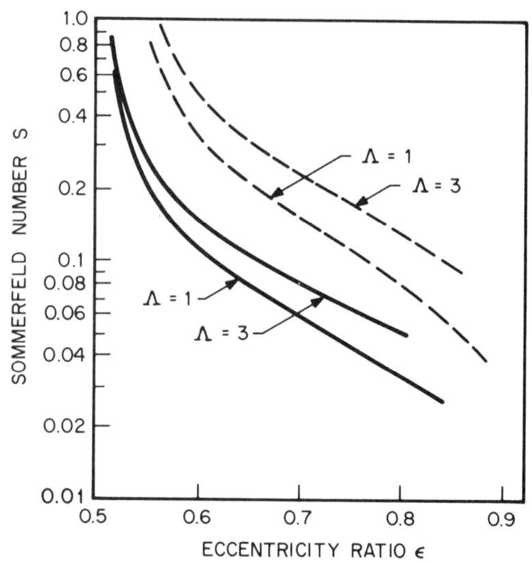

FIG. 4-27 Sommerfeld number versus eccentricity ratio for a noncentrally loaded elliptical gas bearing, $L/D = 1$. *(From Pinkus [4-14].)*

4-8 HYDROSTATIC JOURNAL BEARING DESIGN

Hydrostatic journal bearings (also called *externally pressurized* bearings) offer large radial load-carrying capacities at all rotational speeds (including zero). They exert very little friction and have controllable stiffness. The principal disadvantage of hydrostatic journal bearings is the cost of the pressurized lubricant supply system. These bearings are widely used in the machine-tool industry.

4-8-1 Classification of Bearings and Components

There are basically three types of hydrostatic journal bearings: the single-pad, the multipad, and the multirecess. The various types are depicted in Table 4-21.

The main components of a hydrostatic journal bearing are the pad and the lubricant supply system. The pad consists of a recess, or pocket, region and the surrounding land material. The recess is generally deep compared with the film thickness: typically the depth is taken to be 20 times the radial clearance. In sizing the recess, the land width is reduced as much as practical; however, if the land width is too small, edge effects can become significant. These effects can be avoided if the land width is taken to be greater than 100 times the clearance.

Pressurized lubricant is provided to each bearing recess through the supply system. This system may be as simple as a direct line between the recess and the supply pump. Bearings using such simple systems are termed *noncompensated hydrostatic bearings* and are frequently used to lift highly loaded journals prior to rotation. In that context they are termed *oil lifts.*

In order for a single pump to deliver lubricant to more than one pad or recess, restrictors (or compensating devices) are required in each supply line. Restrictors limit the flow to each pad, thereby permitting all pads to become activated. Without compensating devices pad imbalance would occur. Restrictors, then, are used to control the operation of a hydrostatic bearing.

There are three common types of restrictors: the capillary tube, the orifice plate, and the flow control valve. Table 4-22 compares each device. This table was prepared from a comprehensive series of papers on hydrostatic bearing design by Rippel [4-15]. In addition to flow-load control, the resistor affects the stiffness of the bearing. Bearing stiffness is related to the ability of the bearing to tolerate any changes in the applied load.

4-8-2 Design Parameters

For hydrostatic journal bearings at low rotational speeds, the primary design parameters are maximum load, lubricant flow rate, and stiffness. Of secondary importance are considerations of frictional horsepower and lubricant temperature rise.

The load-carrying capacity of a hydrostatic journal bearing is generally written as

$$W = a_f A_p p_r \tag{4-19}$$

where a_f = pad load coefficient, A_p = projected bearing area, and p_r = recess pressure. The pad load coefficient is dimensionless and physically represents the ratio of the average lubricant pressure in the pad to the lubricant pressure supplied to the pad. O'Donoghue and Rowe [4-16] give equations for a_f for both multirecess and multipad bearings.

The lubricant flow rate may be determined from

$$Q = q_f \frac{h^3 p_r}{\mu} \tag{4-20}$$

TABLE 4-21 Types of Hydrostatic Journal Bearings

Type		Comments
Single-pad journal bearing		1. Included angle less than 180°. 2. Normally used to support unidirectional loads. 3. May have multiple recesses. 4. Lubricant flow and pad design based on a maximum load applied to a uniform film equaling minimum radial clearance $R_B - R_j$.
Multipad journal bearing	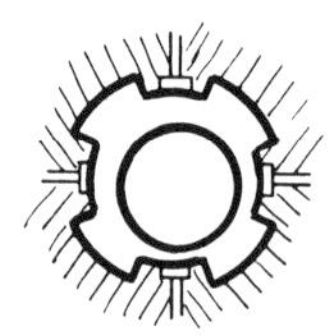	1. Included angle of each pad is less than 180°. 2. Used for rotating, oscillating, or reversing radial loads. 3. Two opposed pads can be used for reversing loads. 4. If angular variation of the load is less than 80°, a two-pad journal bearing may be used. 5. Bearing stiffness is reduced if load is directed between pads. 6. Axial grooves provide cooling capability at high rotational speeds.
Multirecess journal bearing		1. No pressure-reducing grooves between recesses. 2. Usually contains four or more recesses. 3. Used for rotating and reversing loads. 4. Does not require as much flow per recess as multipad type. 5. Circumferential flow from high-pressure pockets to lower-pressure pockets evens pressure distribution around bearing.

where q_f = flow factor for a single pad. The flow factor depends on the geometry and land widths of the bearing.

Journal bearing stiffness expresses the ability of the bearing to accommodate any changes in the applied load. Stiffness is proportional to the slope of the bearing load versus film thickness curve, or

$$s = -\frac{dW}{dh} \tag{4-21}$$

Stiffness will depend on the method of flow control (capillary tube, etc.) and the amount of circumferential flow. O'Donoghue and Rowe [4-16] provide a detailed

TABLE 4-22 Comparison of Three Types of Flow Resistors

	Initial cost	Fabrication and installation costs	Reliability	Availability	Ability to remain unclogged	Serviceability	Life	Flow relationship	Restrictions	Notes
Capillary tube	4†	3	5	5	1	4	5	$Q = k_c \frac{P_s - P_r}{\mu}$ $k_c = \pi D_c^4/128L_c$	$L_c > 20D_c$ $R_e = \frac{4\rho Q}{\pi D_c \mu} < 2000$	1. Laminar flow device. 2. Hypodermic needle tubing can generally be used. 3. To prevent clogging $D_c > 0.025$ in.
Sharp-edged orifice	5	2	5	5	3	4	4	$Q = k_0(P_s - P_r)^{1/2}$ $k_0 = \frac{C_D \pi D_0^2}{2\sqrt{2\rho}}$	$C_D = f(R_e)$ $R_e = \frac{\sqrt{2\rho(P_s - P_r)}D_0}{\mu}$ $C_D \sim 0.6$ for $R_e > 15$ and $D_0/D_{line} < 0.1$	1. Turbulent flow device. 2. To prevent clogging, $D_0 > 0.020$ in.
Flow control valve	1	4	2	5	5	2	2	$Q = k_v$ k_v = constant		1. Relatively expensive. 2. Flow is maintained at constant value.

†Rating numbers: 5 is high (best); 1 is low (worst).

derivation of the journal bearing stiffness and have developed equations for both multipad and multirecess bearings.

4-8-3 Design Procedures

A variety of design procedures are available for hydrostatic bearings. Some are based on experimental findings whereas others are based on numerical solutions of the Reynolds equation.

Many existing methods incorporate numerous charts and tables which clearly place a limit on the methods; in order to use these procedures, the appropriate reference must be consulted. Alternately, O'Donoghue and Rowe [4-16] have developed a general approximate method of design that does not require the use of various design charts. The method is strictly valid for thin land bearings, and many of the parameters are conservatively estimated. The following is a condensation of this design procedure for a multirecess bearing:

DESIGN SPECIFICATION

1. Set the maximum load W.
2. Select the number of recesses n (typically $n = 4$, but $n = 6$ for high-precision bearings).
3. Select a pressure ratio p_r/p_s (a design value of 0.5 is recommended).

BEARING DIMENSIONS

1. Calculate the bearing diameter $D = \sqrt{W/50}$ in.
2. Set width $L = D$.
3. Calculate the axial-flow land width a (refer to Fig. 4-28; recommended value: $a = L/4$).

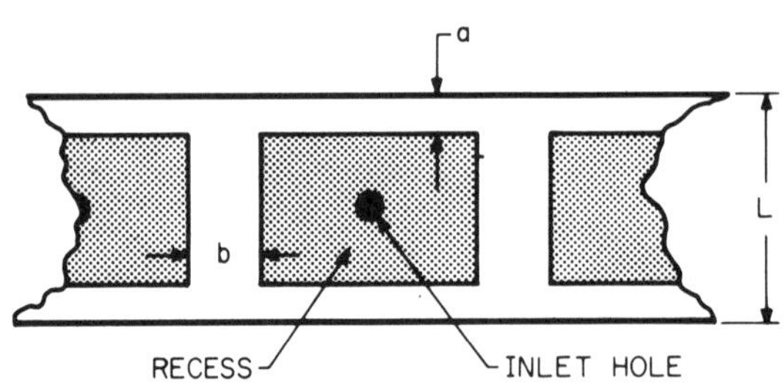

FIG. 4-28 Hydrostatic pad geometry.

4. Calculate the circumferential land width b [refer to Fig. 4-28; recommended value: $b = \pi D/(4n)$].
5. Calculate the projected pad area $A_p = D(L - a)$.
6. Calculate the recess area for one pad only: $A_r = (\pi D - nb)(L - a)/n$.
7. Calculate the effective frictional area: $A_f = (\pi DL/n) - 0.75A_f$.

MISCELLANEOUS COEFFICIENTS AND PARAMETERS

1. Establish a design value of film thickness h_d (recommended value: 50 to 10 times larger than the machinery tolerance on h).

2. Calculate the circumferential flow factor $\gamma = na(L - a)/(\pi Db)$; for recommended values of D, L, a, and b: $\gamma = \frac{3}{4}(n/\pi)^2$.
3. Compute the design dimensionless stiffness parameter $\bar{s}$ from Table 4-23 depending on the method of flow control selected.

TABLE 4-23 Dimensionless Stiffness at Design Condition for Multirecess Bearing

$$\bar{s} = \frac{C_1}{1 + C_{2\gamma}}$$

Number of recesses n	Capillary tube		Sharp-edged orifice		Flow control valve		Recommended value of γ†
	C_1	C_2	C_1	C_2	C_1	C_2	
3	0.27	0.75	0.36	1.0	0.54	1.5	0.68
4	0.96	0.50	1.28	0.67	1.91	1.01	1.22
5	1.03	0.35	1.38	0.46	2.13	0.69	1.90
6	1.08	0.25	1.43	0.33	2.15	0.50	2.74

†SOURCE: Ref. [4-16].

4. Calculate the flow factor $q_f = \pi D/(6an)$.

PERFORMANCE PARAMETERS

1. Calculate the minimum supply pressure $p_{s,\min} = 2W/(a_f A_p)$ [recommended: $a_f = \bar{s}$ (capillary tube at design Table 4-23)].
2. Calculate the stiffness $s = p_s A_p \bar{s}/h_d$ (if this value is too low, either D and/or p_s must be increased).
3. Calculate the minimum running clearance $h_0 = h_d - W/s$.
4. Determine the flow rate $Q = (nq_f p_s h^3 d)/(2\mu)$. A calculated value of viscosity can be used:

$$\mu = \frac{60P_s h^2 d\sqrt{q_f/(2A_f)}}{\pi DN} \qquad \text{reyn}$$

REFERENCES

4-1 G. G. Hirs, "The Load Capacity and Stability Characteristics of Hydrodynamic Grooved Journal Bearings," *ASLE Transactions,* vol. 8, 1965, pp. 296–305.

4-2 E. R. Booser, "Plain-Bearing Materials," *Machine Design,* June 18, 1970, (bearings reference issue), pp. 14–20.

4-3 V. Hopkins, "Self-Lubricating Bearing Materials—A Review," *Assessment of Lubricant Technology,* ASME, 1972, pp. 21–26.

4-4 A. W. J. DeGee, "Selection of Materials for Lubricated Journal Bearings," *Wear,* vol. 36, 1976, pp. 33–61.

4-5 B. R. Reason and I. P. Narang, "Rapid Design and Performance Evaluation of Steady State Journal Bearings—A Technique Amendable to Programmable Hand Calculators," *ASLE Transactions,* vol. 25, no. 4, 1982, pp. 429–449.

4-6 J. Shigley and L. D. Mitchell, *Mechanical Engineering Design,* 4th ed., McGraw-Hill Book Company, 1983, pp. 536–546.

4-7 A. A. Raimondi and J. Boyd, "A Solution for the Finite Journal Bearing and Its Application to Analysis and Design," I, II, and III, *ASLE Transactions,* vol. 1, 1958, pp. 159–174, 175–193, and 194–209, respectively.

4-8 A. Seireg and S. Dandage, "Empirical Design Procedure for the Thermodynamic Behavior of Journal Bearings," *ASME Journal of Lubrication Technology,* vol. 104, April 1982, pp. 135–148.

4-9 H. J. Connors, "An Analysis of the Effect of Lubricant Supply Rate on the Performance of the 360° Journal Bearing," *ASLE Transactions,* vol. 5, 1962, pp. 404–417.

4-10 H. Moes and R. Bosma, "Design Charts for Optimum Bearing Configurations: 1—The Full Journal Bearing," *ASME Journal of Lubrication Technology,* vol. 93, April 1971, pp. 302–306.

4-11 A. A. Raimondi, "A Numerical Solution for the Gas Lubricated Full Journal Bearing of Finite Length," *ASLE Transactions,* vol. 4, 1961, pp. 131–155.

4-12 E. R. Wu, "Gas-Lubricated Porous Bearings of Finite Length—Self-Acting Journal Bearings," *ASME Journal of Lubrication Technology,* vol. 101, July 1979, pp. 338–348.

4-13 V. Castelli and J. Pirvics, "Equilibrium Characteristics of Axial-Grooved Gas-Lubricated Bearings," *ASME Journal of Lubrication Technology,* vol. 89, April 1967, pp. 177–196.

4-14 O. Pinkus, "Analysis of Noncircular Gas Journal Bearings," *ASME Journal of Lubrication Technology,* vol. 87, October 1975, pp. 616–619.

4-15 H. C. Rippel, "Design of Hydrostatic Bearings," *Machine Design,* parts 1 to 10, Aug. 1 to Dec. 5, 1963.

4-16 J. P. O'Donoghue and W. B. Rowe, "Hydrostatic Bearing Design," *Tribology,* vol. 2, Feb. 1969, pp. 25–71.

INDEX